I0051273

Dynamic Simulation of Sodium Cooled Fast Reactors

This book provides the basis of simulating a nuclear plant, in understanding the knowledge of how such simulations help in assuring the safety of the plants, thereby protecting the public from accidents. It provides the reader with an in-depth knowledge about modeling the thermal and flow processes in a fast reactor and gives an idea about the different numerical solution methods. The text highlights the application of the simulation to typical sodium-cooled fast reactor.

The book
- Discusses mathematical modeling of the heat transfer process in a fast reactor cooled by sodium.
- Compares different numerical techniques and brings out the best one for the solution of the models.
- Provides a methodology of validation based on experiments.
- Examines modeling and simulation aspects necessary for the safe design of a fast reactor.
- Emphasizes plant dynamics aspects, which is important for relating the interaction between the components in the heat transport systems.
- Discusses the application of the models to the design of a sodium-cooled fast reactor

It will serve as an ideal reference text for senior undergraduate, graduate students, and academic researchers in the fields of nuclear engineering, mechanical engineering, and power cycle engineering.

Dynamic Simulation of Sodium Cooled Fast Reactors

G. Vaidyanathan

CRC Press
Taylor & Francis Group
Boca Raton London New York

CRC Press is an imprint of the
Taylor & Francis Group, an **informa** business

Cover image: snvv18870020330/Shutterstock

First edition published 2023
by CRC Press
6000 Broken Sound Parkway NW, Suite 300, Boca Raton, FL 33487-2742

and by CRC Press
4 Park Square, Milton Park, Abingdon, Oxon, OX14 4RN

CRC Press is an imprint of Taylor & Francis Group, LLC

© 2023 G. Vaidyanathan

Reasonable efforts have been made to publish reliable data and information, but the author and publisher cannot assume responsibility for the validity of all materials or the consequences of their use. The authors and publishers have attempted to trace the copyright holders of all material reproduced in this publication and apologize to copyright holders if permission to publish in this form has not been obtained. If any copyright material has not been acknowledged please write and let us know so we may rectify in any future reprint.

Except as permitted under U.S. Copyright Law, no part of this book may be reprinted, reproduced, transmitted, or utilized in any form by any electronic, mechanical, or other means, now known or hereafter invented, including photocopying, microfilming, and recording, or in any information storage or retrieval system, without written permission from the publishers.

For permission to photocopy or use material electronically from this work, access www.copyright.com or contact the Copyright Clearance Center, Inc. (CCC), 222 Rosewood Drive, Danvers, MA 01923, 978-750-8400. For works that are not available on CCC please contact mpkbookspermissions@tandf.co.uk

Trademark notice: Product or corporate names may be trademarks or registered trademarks and are used only for identification and explanation without intent to infringe.

ISBN: 978-1-032-25435-7 (hbk)
ISBN: 978-1-032-25437-1 (pbk)
ISBN: 978-1-003-28318-8 (ebk)

DOI: 10.1201/9781003283188

Typeset in Sabon
by SPi Technologies India Pvt Ltd (Straive)

Contents

Preface   xiii
About the Author   xv

1 Introduction   1

  1.1   General 1
  1.2   Basics of Breeding 1
  1.3   Uranium Utilization 2
  1.4   Components of Fast Reactors 5
  1.5   Overview of Fast Reactor Programs 6
  1.6   Need for Dynamic Simulation 9
  1.7   Design Basis 10
  1.8   Plant Protection System 12
  1.9   Sensors and Response Time 14
  1.10 Scope of Dynamic Studies 15
  1.11 Modeling Development 16
  Assignment 17
  References 17

2 Description of Fast Reactors   19

  2.1   Introduction 19
  2.2   Fast Breeder Test Reactor (FBTR) 19
       2.2.1   Reactor Core 21
       2.2.2   Reactor Assembly 23
       2.2.3   Sodium Systems 24
       2.2.4   Decay Heat Removal 27
       2.2.5   Generating Plant 28
       2.2.6   Instrumentation and Control 28
       2.2.7   Safety 29

2.3   Prototype Fast Breeder Reactor 31
    2.3.1   Reactor Core 32
    2.3.2   Reactor Assembly 33
    2.3.3   Main Heat Transport System 35
    2.3.4   Steam Water System 35
    2.3.5   Instrumentation and Control 36
    2.3.6   Safety 37
2.4   Neutronic Characteristics of SFRs 38
2.5   Thermal-Hydraulic Characteristics of SFR 40
Assignment 41
References 41

3 Reactor Heat Transfer   **43**

3.1   Introduction 43
3.2   Reactor Core 43
    3.2.1   Core Description 44
    3.2.2   Fuel Pin 45
    3.2.3   Subassembly 47
3.3   Coolant Selection 47
3.4   Control Material Selection 48
3.5   Structural Material Selection 48
3.6   Heat Generation 49
3.7   Reactivity Feedback 51
    3.7.1   Doppler Effect 52
    3.7.2   Sodium Density and Void Effects 53
    3.7.3   Fuel Axial Expansion Effect 53
    3.7.4   Structural Expansion 54
    3.7.5   Bowing 54
3.8   Decay Heat 56
3.9   Solution Methods 57
    3.9.1   Prompt Jump Approximation 57
    3.9.2   Runge Kutta Method 58
    3.9.3   Kaganove Method 58
    3.9.4   Comparison of the Different Methods 58
    3.9.5   Solution Methodology 59
3.10 Heat Transfer in Primary System 61
    3.10.1 Core Thermal Model 61
    3.10.2 Fuel Restructuring 62
    3.10.3 Gap Conductance 63
    3.10.4 Fuel Thermal Model 63
    3.10.5 Solution Technique 64

3.11 Determination of Peak Temperatures: Hot Spot Analysis 64
3.12 Core Thermal Model Validation in FBTR and SUPER
      PHENIX 65
3.13 Mixing of Coolant Streams in Upper Plenum 66
      3.13.1 Solution Technique 69
3.14 Lower Plenum/Cold Pool 70
3.15 Grid Plate 74
3.16 Heat Transfer Correlations for Fuel Rod Bundle 75
Assignment 76
References 77

4 IHX Thermal Model   **79**

4.1 Introduction 79
4.2 Experience in PHENIX 79
4.3 Thermal Model 82
4.4 Solution Techniques 83
     4.4.1 Nodal Heat Balance Scheme 83
     4.4.2 Finite Differencing Scheme 84
4.5 Choice of Numerical Scheme 85
     4.5.1 Nodal Heat Balance for Unbalanced Flows 85
     4.5.2 Modified Nodal Heat Balance Scheme (MNHB) 86
4.6 Heat Transfer Correlations 89
4.7 Validation 90
Assignment 91
References 92

5 Thermal Model of Piping   **95**

5.1 Introduction 95
5.2 Thermal Model 96
5.3 Solution Methods 97
5.4 Comparison of Piping Models 98
Assignment 100
References 100

6 Sodium Pump   **101**

6.1 Introduction 101
6.2 Electromagnetic Pumps 101
6.3 Centrifugal Pump 102
     6.3.1 Pump Hydraulic Model 104
     6.3.2 Pump Dynamic Model 104

6.3.3   *Pump Thermal Model 108*
Assignment 109
References 109

7 Transient Hydraulics Simulation   111

7.1   *Introduction 111*
7.2   *Momentum Equations 111*
7.3   *Free Level Equations 113*
7.4   *Core Coolant Flow Distribution 113*
7.5   *IHX Pressure Drop Correlations 116*
    7.5.1   *Resistance Coefficient for Cross-Flow 117*
    7.5.2   *Resistance Coefficient for Axial Flow 118*
7.6   *Pump Characteristics 118*
7.7   *Computational Model 118*
7.8   *Validation Studies 119*
7.9   *Secondary Circuit Hydraulics 120*
    7.9.1   *Secondary Hydraulics Model 121*
    7.9.2   *Natural Convection Flow in Sodium Validation
        Studies 122*
Assignment 123
References 123

8 Steam Generator   125

8.1   *Introduction 125*
8.2   *Heat Transfer Mechanisms 125*
8.3   *Steam Generator Designs 127*
    8.3.1   *Conventional Fossil-Fueled Boilers 127*
        8.3.1.1   *Drum Type 127*
        8.3.1.2   *Once Through Steam Generators
            (OTSG) 128*
    8.3.2   *Sodium-Heated Steam Generators 129*
8.4   *Thermodynamic Models 133*
8.5   *Mathematical Model 138*
8.6   *Heat Transfer Correlations 139*
    8.6.1   *Single-Phase Liquid Region 139*
    8.6.2   *Nucleate Boiling 139*
    8.6.3   *Dry-Out 141*
    8.6.4   *Post Dry-Out 142*
    8.6.5   *Superheated Region 142*
    8.6.6   *Sodium Side Heat Transfer 142*

8.7   *Pressure Drop 142*
8.8   *Computational Model 144*
    8.8.1   *Solution of Water/Steam Side Equations 144*
    8.8.2   *Solution of Sodium, Shell, and Tube Wall Equations 145*
8.9   *Steam Generator Model Validation 146*
Assignment 149
References 149

9 Computer Code Development   151

9.1   *Introduction 151*
9.2   *Organization of DYNAM 151*
9.3   *Axisymmetric Code STITH-2D 154*
9.4   *One-dimensional CFD-coupled Dynamics Tool 155*
9.5   *Comparison of Predictions of DYANA-P and DYANA-HM 156*
References 162

10 Specifying Sodium Pumps Coast-Down Time   163

10.1   *Introduction 163*
10.2   *Impact of Coast-Down Time in Loop-Type SFR 163*
10.3   *Impact of Coast-Down Time in Pool-Type SFR 165*
10.4   *Considerations for Determining Flow Coast-Down Time 166*
10.5   *SCRAM Threshold versus Coast-Down Time 169*
    10.5.1   *FHT Effect on Maximum Temperatures 170*
    10.5.2   *FHT to Avoid SCRAM for Short Power Failure 171*
10.6   *Secondary Pump FHT 171*
10.7   *Primary FHT for Unprotected Loss of Flow 172*
Assignment 174
References 174

11 Plant Protection System   177

11.1   *Introduction 177*
11.2   *Limiting Safety System Settings (LSSS) for FBTR 177*
    11.2.1   *Safety Signals and Settings 178*
    11.2.2   *Limiting Safety System Settings (LSSS) Adequacy 178*
11.3   *Limiting Safety System Settings for PFBR 180*
    11.3.1   *Design Basis Events 181*
    11.3.2   *Core Design Safety Limits 181*
    11.3.3   *Selection of SCRAM Parameters 182*

11.4   Shutdown System 182
11.5   Event Analysis 184
Assignment 187
References 187

12 Decay Heat Removal System   189

12.1   Introduction 189
12.2   Natural Convection Basics 189
12.3   DHR System Options in SFR 191
        12.3.1   DHR in Primary Sodium 191
        12.3.2   DHR in Secondary Sodium 191
        12.3.3   Steam Generator Auxiliary Cooling System 192
        12.3.4   DHR Through Steam Water System 192
        12.3.5   Reactor Vessel Auxiliary Cooling System 193
12.4   DHR in FBTR 194
        12.4.1   Heat Removal by Air in SG Casing 195
        12.4.2   Loss of Off-site and On-site Power with SG Air
                 Cooling 197
        12.4.3   Loss of Off-site and On-site Power without
                 Reactor Trip (ULOF) 200
12.5   DHR in PFBR 202
        12.5.1   Thermal Model 204
        12.5.2   Decay Heat Exchanger (DHX) Thermal Model 204
        12.5.3   Hot Pool Thermal Model 205
        12.5.4   Air Heat Exchanger (AHX) Thermal Model 207
        12.5.5   Piping 209
        12.5.6   Expansion Tank 209
        12.5.7   Air Stack/Chimney 209
        12.5.8   Hydraulic Model of SGDHR 210
        12.5.9   DHDYN Validation on SADHANA Loop 211
12.6   Role of Inter-Wrapper Flow 212
12.7   Role of Secondary Thermal Capacity 213
Assignment 214
References 214

13 Modeling of Large Sodium–Water Reaction   217

13.1   Introduction 217
13.2   Leak Rate 218
        13.2.1   Water Leak Rate Model 218
        13.2.2   Steam Leak Rate Model 220

13.3   Reaction Site Dynamics 221
        13.3.1   Spherical Bubble Model 223
        13.3.2   Columnar Bubble Model 223
        13.3.3   Solution Technique 224
        13.3.4   Validation of Reaction Site Model 224
13.4   Sodium Side System Transient 226
13.5   Discharge Circuit System Transient 226
13.6   Analysis of Pressure Transients for PFBR 227
13.7   Failure of a Greater Number of Tubes Than Design Basis
        Leak 228
Nomenclature 229
Assignment 230
References 230

Appendix A Brief Description of Codes   233
Appendix B Primary Pump Discharge Pipe
            Break Modeling   245
Index   253

Preface

There are nearly 440 electricity-generating nuclear reactors that are in operation worldwide. Most of these are light water reactors that use U235 enriched upto 5%, with a smaller number of heavy water reactors which use 0.7% U235 present in natural uranium. Since water reactors utilize only the U235 portion that is fissile, the full fuel (U235 + U238) utilization is possible if we could utilize the remaining U238. U238 is not a fissionable material like U235. However, it gets converted to Pu239 when it gets bombarded by neutrons that have not taken part in the fission of U235. This conversion ratio is a strong function of the neutron spectrum. In water-cooled reactors the neutron energy is brought down to low levels to have greater probability of fission using a moderator. However, the ratio of fission to absorption probability is much higher when the neutron energy levels are high, and this results in greater amounts of neutrons being available for converting U238 to Pu 239. This is possible only in fast reactors. If we are to effectively utilize the natural uranium resources and convert unutilized U238 to fissile material, then fast reactors have an important role. In addition, in the fast neutron spectrum, many long half-life actinides are burned and converted to short-life elements, thus reducing the life span of the waste produced in water reactors. Based on this, few countries have embarked on the construction of fast reactors cooled by sodium. Furthermore, there has been a slowdown in fast reactor development due to findings of new natural uranium resources. Meanwhile, new emerging concepts have once again increased the relevancy of sodium-cooled fast reactors (SFR). Half of the new-concepts, or fourth-generation, nuclear reactors are fast reactors.

Though the overall record of clean and safe power from nuclear reactors is good, steps to improve safety and availability are ongoing. To achieve this goal, the physical phenomena inside the reactor must be carefully and sufficiently identified and investigated. Conducting experiments is a good way to investigate these problems. However, for a nuclear power plant, the experiment can hardly cover everything because of the high cost and measurement difficulties. Simulation represents an attractive alternative thanks to its cost efficiency, easy repeatability, and transparency of results.

The design of a SFR includes determination of the transient thermal loading seen by the different components under various transients. There is also the need to ensure that the safety limits on component temperatures are not exceeded and reactor is shut down automatically. It is also necessary to analyze the limits on the operating parameters that will shut down the reactor. With the licensing need for operator training on simulator platforms, a dynamic simulation tool assumes more importance.

The author has taught the subject of fast reactor design to university students for nearly a decade. During this he felt the need for a book that brings out the simulation models of fast reactor components in a simple manner for better understanding by the students. Such a book is also necessary for engineers joining the nuclear establishments, in developing dynamic simulation tools for design analysis and training simulators. There have been many books that describe fast reactors, but very few are devoted completely to the transient simulation. This book aims to fulfil this gap.

Chapter 1 gives an insight into the fast reactor characteristics and programs in different countries. Chapters 2 through 8 deal with the thermal hydraulic simulation of the different components, namely reactor core, intermediate heat exchanger, primary and secondary piping, and steam generator. Chapter 9 deals with code organization while Chapters 10 and 11 bring out the application of simulation to give inputs for sodium pump coast downtime and reactor protection system. Chapters 12 and 13 take the reader through modeling of the decay heat removal system and modeling of sodium water reaction in the steam generator, respectively. This book is adaptable for a single-semester course for students of nuclear science and engineering.

The author acknowledges with thanks the support and inputs from several former colleagues: Dr. K. Velusamy, Mr. N. Kasinathan, Dr. K. Natesan, and Dr. U. Parthasarathy of the Indira Gandhi Centre of Atomic Research, Kalpakkam, India. The author wishes to thank his grandson Master. Atharv Shivakumar, who typed out all the equations. The author also gratefully acknowledges the mentoring and motivation provided for the simulation activities by Mr. S. R. Paranjpe (deceased), former director of India Gandhi Centre for Atomic Research, Kalpakkam, fondly known as the father of fast reactors in India.

Dr. G. Vaidyanathan

About the Author

Dr G. Vaidyanathan, BE, MBA, PhD, had served the Department of Atomic Energy, India for 38 years until 2010, reaching the post of a director of the Fast Reactor Technology Group at the Indira Gandhi Centre for Atomic Research (IGCAR), Kalpakkam, which is devoted to the development of fast reactors in India. He is a specialist in numerical and experimental thermal hydraulics and safety analysis. He was one of the key persons involved in the design and development of the experimental fast breeder test reactor (FBTR) and the prototype fast breeder reactor (PFBR) in India. He has contributed to the in-house development of thermal hydraulic computer codes, and during commissioning tests in FBTR the predictions of these codes were substantiated.

After 2010, Dr. Vaidyanathan has been teaching nuclear energy and alternative systems at various Indian universities. He has published four books on nuclear energy, which have been adapted by many professors teaching nuclear subjects in universities in India and elsewhere. He has developed a video module comprising 30 lectures on nuclear reactors and safety under the NPTEL program of IIT & IISc, and these have been extensively used.

Dr. Vaidyanathan is a lifetime fellow of the Institution of Engineers India and a lifetime member of the Indian Nuclear Society and Indian Society of Heat & Mass Transfer. He has 37 journal publications to his credit. He continues to contribute to the department as a member of the sodium safety panel (SSP) at IGCAR and the Advisory Committees for Safety Review of various projects (ACPSR) at the Atomic Energy Regulatory Board, India.

Chapter 1

Introduction

1.1 GENERAL

The history of fast reactors goes back to the construction of an experimental fast reactor, Clementine, USA, in 1946 (IAEA, 2013). The coolant was liquid mercury. This reactor used plutonium as fuel and was built to demonstrate the feasibility of operating a fast reactor. Next was the construction and operation of Experimental Breeder Reactor-I (EBR-I) again in USA in 1951. This used enriched uranium 235 as fuel and was cooled by a liquid sodium-potassium alloy (NaK). It is notable that EBR-I was the first nuclear reactor to generate electrical power on December 20, 1951. Subsequently many fast reactors using sodium as coolant have been built and operated in many countries, with a cumulative international operational experience of more than 400 years. This chapter focuses on the need for breeder reactors and gives an overview of the fast reactor programs around the world. It also brings out the need for dynamic simulation toward plant design and assessment of transient response to various design basis events.

1.2 BASICS OF BREEDING

The discovery of nuclear fission in the 1930s, followed by many reactor experiments, indicated that the isotopes U_{233}, U_{235}, and Pu_{239} are fissionable when bombarded by neutrons with energies in the low (<1 eV) and intermediate ranges (around 1 MeV) (Waltar et al., 2012). The low-energy neutrons are also referred to as thermal neutrons, and reactors in which fission is based on thermal neutrons, like the light-water and heavy-water reactors, are called thermal reactors.

Of the fissionable isotopes only U_{235} is available in nature. Natural uranium contains ~0.7% of U_{235} and the rest is U_{238}. U_{238} and Th_{232} capture neutrons at energies below 1 MeV range and get converted to Pu_{239} and U_{233}, respectively. Pressurized water reactors (PWR) and boiling water reactors (BWR) use enriched U_{235} (~2 to 5% U_{235}) compared to natural uranium (~0.7% U_{235}) in pressurized heavy water reactors (PHWR). Since thermal

DOI: 10.1201/9781003283188-1

reactors utilize only the U_{235}, which is fissile, the full fuel (U_{235} + U_{238}) utilization is possible if the remaining U_{238} can be utilized.

If more fissile isotopes could be produced from U_{238} and Th_{232}, referred to as fertile isotopes, than were destroyed in the fission chain reaction, it would be possible to effectively utilize the fertile isotopes. This process of producing more fuel than what is consumed is referred to as breeding. It was soon learned that the ratio of number of neutrons emitted in fission per neutron absorbed, called ETA (η) for fissile isotope Pu_{239}, is higher in a fast neutron spectrum than in a thermal energy spectrum as in water reactors (Figure 1.1). Higher number of emitted neutrons meant more neutrons available for conversion of U_{238} to Pu_{239}.

Hence the idea of a breeder reactor operating on fast neutrons was born to effectively utilize the U_{238}. It was also found possible to breed with thermal reactors, and the Th_{232}-U_{233} cycle was determined to be a better cycle for thermal breeding.

1.3 URANIUM UTILIZATION

Natural uranium contains about 0.7% U_{235} and 99.3% U_{238}. In any reactor some of the U_{238} component is turned into several isotopes of plutonium during its operation. Two of these, Pu_{239} and Pu_{241}, then undergo fission in the same way as U_{235} to produce heat. In a fast neutron reactor (FNR) this process can be optimized so that it "breeds" fuel. Some U_{238} is also burned directly with neutron energies above 1 MeV (fast fission).

The FNR has no moderator and relies on fast neutrons alone to cause fission, which for uranium is less efficient than using slow neutrons. FNRs usually use plutonium as the basic fuel, since the number of neutrons produced per Pu_{239} in fast neutron fission is 25% more than from uranium, and this means that there are enough neutrons (after losses) not only to maintain the chain reaction but also continually to convert U_{238} into more Pu_{239} (Figure 1.2). The coolant is a liquid metal (normally sodium) to avoid any neutron moderation and provide a very efficient heat transfer medium. So, the FNR "burns" and "breeds" fissile plutonium. Both U_{238} and Pu_{240} are "fertile" materials, i.e., by capturing a neutron they become (directly or indirectly) fissile Pu_{239} and Pu_{241}, respectively. While the conversion ratio (the ratio of new fissile nuclei to fissioned nuclei) in a thermal reactor is around 0.6, that in a FNR may exceed 1.0.

The conventional FNRs, also referred to as sodium-cooled fast reactors (SFR), built so far have a "fertile blanket" of depleted uranium (U_{238}) around the core, and this is where much of the Pu_{239} is produced. Neutron activity is very low in the blanket, so the plutonium produced there remains almost pure Pu_{239} – largely not burned or changed to Pu_{240}. The blanket can then be reprocessed (as is the core) and the plutonium recovered for use in the core, or for further SFRs.

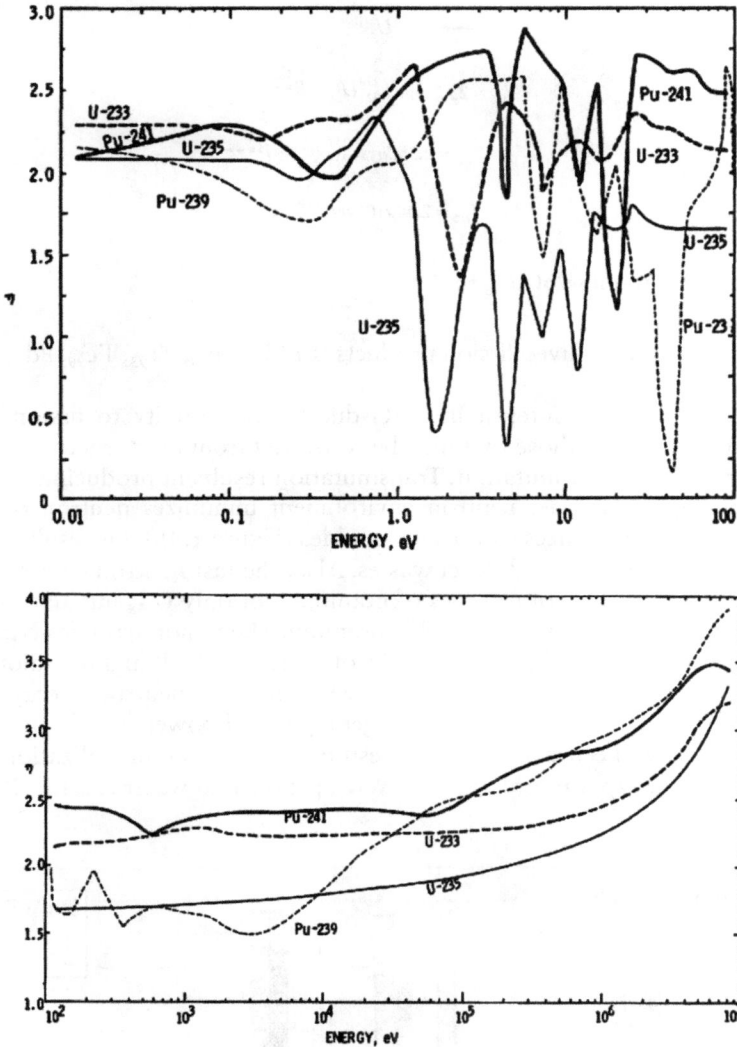

Figure 1.1 η values of fissile isotopes: (a) neutron energy 0.01 to 100ev, (b) neutron energy 10^2 ev to 10^7 ev.

Source: ERRI NP-359, Assessment of Thorium Fuel Cycles in Pressurized Water Reactors, Final Report, February 1977.

After recovering U and Pu from spent nuclear fuel of thermal reactors (PWR, BWR, PHWR) by conventional reprocessing, the disposal of high-level radioactive wastes is a major concern in many countries. Most of the radioactive hazard remaining in high-level radioactive wastes after thousands of years comes from minor actinides (MA); isotopes of Np, Am, and

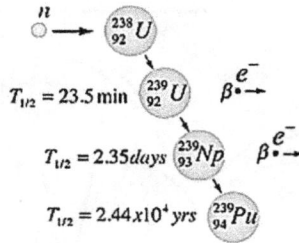

Figure 1.2 Conversion of U_{238} to Pu_{239}.

Cm, and some long-lived fission products (LLFPs; Se_{79}, Zr_{93}, Tc_{99}, Pd_{107}, I_{129}, and Cs_{135}).

There is renewed interest in SFRs due to their ability to fission these actinides, including those that may be recovered from used reactor fuel and is referred to as transmutation. Transmutation results in producing shorter-lived isotopes. The fast neutron environment minimizes neutron capture reactions and maximizes fissions in actinides (Figure 1.3). This results in less long-lived nuclides in high-level wastes. Also, the fast neutron environment is required for fission of isotopes of uranium, not only U_{238} but also others, which may be significant in recycled uranium. The minor actinides Np, Am, and Cm (as well as the higher isotopes of plutonium), all highly radiotoxic, are much more readily destroyed by fission in a fast neutron energy spectrum, and they also contribute to the generation of power.

Studies have been carried out to estimate the uranium utilization in a pressurized heavy water reactor (PHWR), pressurized water reactor (PWR),

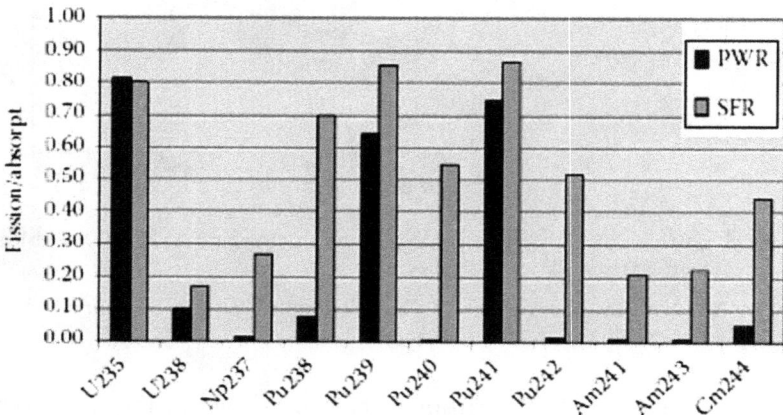

Figure 1.3 Comparison of fissionability fraction in thermal and fast reactors.

Source: Wade, D. C. and Hill, R. N., The design rationale of the IFR, *Prog. Nucl. Energy* 31, No. 1–2, 1997.

and a similar-power SFR (Raj et al., 2015). A 1,000 MWe PHWR operating with a thermodynamic efficiency of 28% would produce a thermal power of 3,600 MWt. The average fission energy release per unit mass of fuel, also referred to as burn-up for PHWRs, is around 7,000 MWd/ton. For this typical fuel burn-up with a load factor of 95%, the annual natural uranium requirement works out to ~178 tons. The irradiated fuel discharged contains ~98.91% depleted uranium (DU), 0.7% fission products, 0.385% Pu, and 0.005% MA. Hence, the effective utilization of natural uranium is (0.7 + 0.385), that is, equal to ~1.1%. Similar estimates were made for PWR and SFR without recycling (open cycle), and it was found that effective utilization was ~5% in a PWR and ~22% in a SFR. Recycling involves recovering U and Pu from the spent fuel, refabricating, and reloading back into the reactor (closed cycle). With recycling, the utilization in fast reactors would be still better and waste can be reduced further.

1.4 COMPONENTS OF FAST REACTORS

Simplified flow sheets of loop-type and pool-type SFRs are shown in Figures 1.4 and 1.5. In the loop-type concept, the primary sodium system components, namely reactor core, heat exchanger, and pump, are kept in separate vessels and connected through pipes, while in the pool type all the components are in a single vessel as shown. Both concepts can be engineered to meet the safety requirements. In both concepts the heat generated by nuclear fission in the core is transported by primary sodium to the

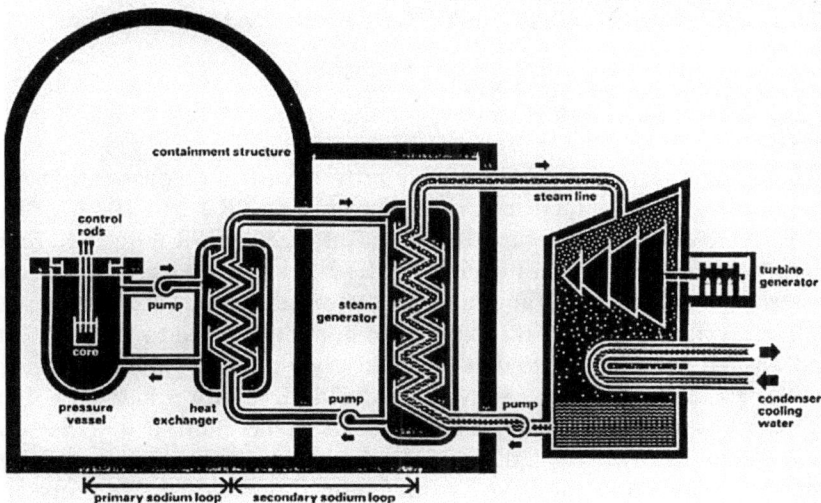

Figure 1.4 Schematic of a loop-type fast reactor.

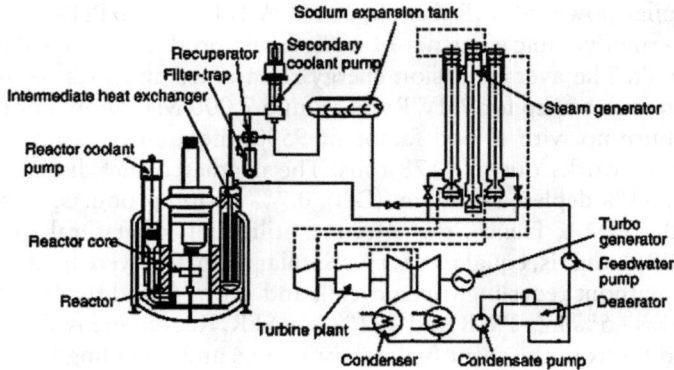

Figure 1.5 Schematic of a pool-type fast reactor.

intermediate heat exchanger (IHX), where it transfers heat to nonradioactive secondary sodium (Waltar et al., 2012). The secondary sodium in turn transfers heat to water in the steam generator (SG) to produce superheated steam to run a turbine generator and produce electricity. All sodium components, namely reactor vessel, IHX, and pump, have a free level topped up by argon as cover gas. The presence of cover gas ensures that sodium does not leak out through the top clearances and result in a sodium fire. The steam cycle is similar in both concepts. In some experimental reactors, the SG is replaced by a terminal air exchanger cooled by air blowers and heat is ejected to atmosphere.

1.5 OVERVIEW OF FAST REACTOR PROGRAMS

Many experimental SFRs have been built to date (Table 1.1). Clementine reactor in the USA was the first one, and the latest was the CEFR reactor in China.

Though the feasibility of SFRs was proved with the construction and operation of experimental reactors like Clementine, EBR-I, EBR II, LAMPRE, and Fermi in the USA, BR-5 and BR-10 in Russia, and DFR in the UK, there was no effort to go for commercial SFRs until the 1960s, when it was assessed that the natural uranium resources would not be sufficient to proceed with water reactors. Also, the cost of uranium started to increase. The need was felt to go in for breeders to effectively utilize the natural U_{238} present. The French had built the 40-MWt RAPSODIE reactor, and the Russians had operated the BOR-60 reactor with power generation in the 1960s. A 250-MWe power reactor PHENIX was commissioned by France in 1973. Around the same time the British commissioned their 250-MWe prototype fast reactor (PFR). The Russians built the BN 350 plant in Kazakhstan. The uniqueness of this plant was that part of its steam generation (~150MWe)

Table 1.1 Experimental fast reactors

Reactor	Country	Year of Criticality	Thermal Rating (MW)	Power Rating (MW)	Fuel/Coolant
CLEMENTINE	US	1946	0.025		Pu Metal/Hg
EBR-I	US	1951	1.2	0.2	U Metal/NaK
BR-1/2	USSR	1956	0.1	-	Pu Metal/Hg
BR-5/10	USSR	1958	5/10	-	PuO_2, UC/Na
Dounreay (DFR)	UK	1959	60	15	U Metal/NaK
LAMPRE	US	1961	1	-	Liq. Pu/Na
Fermi (EFFBR)	US	1963	200	65	U Metal/Na
EBR-II	US	1963	62	20	U Metal/Na
Rapsodie	France	1967	40	-	$U-PuO_2$ /Na
SEFOR	US	1969	20	-	$U-PuO_2$/Na
BOR-60	USSR	1969	60	12	UO_2/Na
KNK-2	Germany	1977	58	21	UO_2/Na
JOYO	Japan	1977	100	-	$U-PuO_2$ /Na
FFTF	US	1980	400	-	$U-PuO_2$/Na
FBTR	India	1985	40	13	U-PuC/Na
CEFR	China	2010	65	20	$U-PuO_2$/Na

Hg: mercury; Na: sodium; NaK: sodium-potassium alloy

was utilized for water desalination. The Germans and Japanese started with their experimental fast reactors KNK and Joyo, respectively, at this time. The USA built a large experimental reactor for testing fuels called the fast flux test facility (FFTF). Then came the BN-600 plant in Russia, followed by a 1200-MWe SUPER PHENIX reactor in France. SUPER PHENIX was built by a consortium of companies from France, Germany, and Italy. The 40-MW fast breeder test reactor (FBTR) was commissioned in 1985 in India. Unfortunately, two reactors, CRBRP in USA and SNR300 in Germany, were built but did not become operational due to political opposition. The SUPER PHENIX reactor plant was shut down after 12 years of operation due to some technical issues and negative political pressure. Japan built and operated a 300-MWe power reactor MONJU. However, due to a sodium leak in the secondary loop after which the plant was shut down, and with the Fukushima accident, the license to restart has been denied. The 1500-MWe European fast reactor (EFR) was designed by France, the UK, and Germany jointly in the 1990s. However, the lack of demand for new power resources and the finding of new natural uranium sources led to the slowing down of the SFR programs in these countries. China started its fast reactor program with the Chinese experimental fast reactor (CEFR) in 2010. Russia commissioned the BN-800 reactor in 2014. Prototype fast reactors built to date are presented in Table 1.2.

Table 1.2 Prototype fast reactors

Reactor	Country	Critical	MWe/ MWt	Fuel	Coolant	Layout
BN-350	USSR	1972	150 /1000	UO_2	Na	Loop
PHENIX	France	1973	250/ 568	UO_2-PuO_2	Na	Pool
PFR	UK	1974	250/ 600	UO_2-PuO_2	Na	Pool
BN-600	USSR	1980	600 /1470	UO_2	Na	Pool
BN-800	USSR	2014	800/ 2100	UO_2	Na	Pool
S. PHENIX	France	1983	1200/ 3000	UO_2-PuO_2	Na	Pool
SNR-300	Germany	----	327/ 770	UO_2-PuO_2	Na	Loop
MONJU	Japan	1987	300 /714	UO_2-PuO_2	Na	Loop
CRBRP	US	---	375/ 975	UO_2-PuO_2	Na	Loop

The impact of greenhouse gases emitted by coal plants and the consequent effects of global warming, acid rain, etc. have led to renewed interest in nuclear power in developed countries since the 1990s. This resulted in different countries coming together to focus on the types of reactors that must be built in the future, called fourth-generation, or GEN-IV, reactors (Bahman, 2020). It is to be noted that in the GEN-IV, fast reactors are one of the potential candidates favored by all countries (Table 1.3). The countries pursuing the fast reactor program presently are Japan, Russia, India, Korea, China, and France. In India there is vigorous attention being paid to

Table 1.3 GEN-IV systems

System	Neutron Spectrum	Coolant	Temp. ° C	Fuel Cycle	Power MWe
Very High Temperature Reactor (VHTR)	Thermal	Helium	900–1000	Open	250–300
Sodium Fast Reactor (SFR)	Fast	Sodium	550	Closed	30–150, 300–1,500, 1,000–2,000
Supercritical Water-Cooled Reactor (SCWR)	Thermal Fast	Water	510–625	Open Closed	300–700 1,000–1,500
Gas-Cooled Fast Reactor (GFR)	Fast	Helium	850	Closed	1,200
Lead-Cooled Fast Reactor (LFR)	Fast	Lead	480–800	Closed	180–210, 300–1,200, 600–1,000
Molten Salt Reactor (MSR)	Epithermal	Fluoride Salts	700–800	Closed	1,000

breeder reactors essentially due to absence of sufficient natural uranium sources, constraints on getting the same on the world market, and availability of abundant sources of thorium. Effective utilization of natural uranium sources is possible only with fast reactors.

1.6 NEED FOR DYNAMIC SIMULATION

Dynamic simulation is one of the important analysis domains that play a significant role from conception to decommissioning of a nuclear power plant. This is important in arriving at a suitable design of a plant protection system, demonstration of plant safety, and establishment of safe plant operation. With the need to train operators before working on the actual plant, computer-based simulators are needed. The simulator is an interactive tool for teaching and operator training of the basics of the reactor operation, reactor physics, and thermal hydraulics. It also needs a dynamic simulation software.

These simulations are carried out using system-level thermal hydraulic computer codes (known as plant dynamic codes) that predict the neutronic/thermal hydraulic behaviors of an entire plant. Plant dynamic codes have computational models to represent various phenomena involved in normal operation as well as transient conditions involving reactor physics, coolant flow, and heat transport in various coupled loops and components of the plant. Dynamic thermal hydraulic simulation of the entire SFR system is required in the design evolution to provide thermal loads for structural design, identify the process parameters for initiating safety actions, arrive at the permissible response time of instruments measuring these parameters, establish operating specifications, as well as in safety analysis of the plant. The degree of interdependence of a component on other components can be assessed only after developing a tool (analytical, empirical, or semi-empirical) that encompasses all essential components.

The degree of model sophistication for the components is dependent on the problems under investigation. As an example, for safety-related transients, the decay heat removal system and heat-generating core need to be modeled in sufficient detail, while the balance of the plant can be modeled in less detail. Dynamic simulation of the entire plant is essential for safety demonstration and licensing. The impact of day-to-day transients needs to be assessed individually to ascertain that temperatures and pressures in the entire systems are within the design limits. In case they cross the limits, automatic safety actions need to be added in the safety system design to ensure that all parameters remain within limits. While a single event may not pose any threat to structural integrity, cumulative impact due to all events can cause failure due to creep and fatigue. All structural components need to be evaluated for the cumulative damage due to all transients over the period of their design life. The whole spectrum of conceivable events in each plant design must be identified and classified.

1.7 DESIGN BASIS

Classification of events in a nuclear power reactor is prompted by the necessity of approaching safety in a systematic manner. Overall safety approach is based on the defense-in-depth philosophy to recognize five levels of design (IAEA, 1996). Successful defense in depth requires creating, maintaining, and updating multiple independent and redundant layers of protection to compensate for potential human and mechanical failures so that no single layer, no matter how robust, is exclusively relied upon (Table 1.4).

The first level provides adequate and reliable functional design to provide a quality product relatively fault free. The second level renders protection against equipment or human failures through a reliable and comprehensive protection system and redundant heat removal systems. The third level develops additional protection by considering extremely unlikely faults in the design basis. The fourth level deals with mitigating fault progression and accident management. The fifth level deals with emergency response.

Postulated initiating events (PIEs) include anticipated operational occurrences or accident conditions, including equipment failure, human errors, and external events (natural or human induced). The postulated events are

Table 1.4 Levels of defense in depth (IAEA, 1996)

Level	Objective	Defense/Barrier
Level 1	Prevention of abnormal operation and failures by design	Conservative design, construction, maintenance, and operation in accordance with appropriate safety margins, engineering practices, and quality levels
Level 2	Prevention and control of abnormal operation and detection of failures	Control, indication, alarm systems or other systems and operating procedures to prevent or minimize damage from failures
Level 3	Control of faults within the design basis to protect against escalation to an accident	Engineering safety features, multiple barriers, and accident or fault control procedures
Level 4	Control of severe plant conditions in which the design basis might be exceeded, including protecting against further fault escalation and mitigation of the consequences of severe accidents	Additional measures and procedures to protect against or mitigate fault progression and for accident management
Level 5	Mitigation of radiological consequences of significant release of radioactive materials	Emergency control and on- and off-site emergency response

grouped together according to their frequency of occurrence and systems designed such that the probability of any event is considered acceptable in terms of damage to the plant. The design basis events (DBE) are classified into four categories according to their anticipated frequency of occurrence and potential consequences (IAEA, 2019).

1. Normal Conditions: This includes plant start-up, plant shutdown, load changing, and operational transients. Here the plant accommodates within the fuel and plant operating margins without the need for automatic or manual protective action.
2. Anticipated Operational Occurrences: This includes events of moderate frequency like loss of a single pump, loss of off-site power, inadvertent withdrawal of a control rod, turbine trip, sudden closure of a turbine isolation valve, etc. These events may need a protective action to trip the reactor and restart after corrective action.
3. Design Basis Accidents: These include infrequent events like a single pump seizure, station blackout (loss of off-site and on-site power), malfunction in the operation of a reactor plant controller, etc. These events could result in small failure of fuel clad (hereafter mostly referred to as just clad) or plant components, to preclude resumption of the plant operation for a considerable time.
4. Design Extension Conditions: These include postulated events of very low frequency like loss of reactor trip function following upset or emergency events, steam line rupture, large reactivity insertion, etc. These events can result in substantial fuel and/or component failures and radioactivity release within specified regulatory limits.

The design basis events of some of the reactors is presented in a NUREG report (Agrawal, 1983). Despite some differences due to licensing decisions, there appears to be a considerable degree of unanimity in the selection (definition) of design basis events in all of these plants. The design basis events commonly analyzed are given below (Kasinathan et al., 2002). These are the enveloping events that are needed to demonstrate safety.

- One primary pump trip (Category 2)
- One primary pump seizure (Category 3)
- One secondary pump trip (Category 2)
- One secondary pump seizure (Category 3)
- Off-site power failure with emergency backup (Category 2)
- Short-term (<4 h) off-site power failure and emergency backup failure (Category 3)
- Long-term (>4 h) off-site power failure and emergency backup failure (Category 4)
- Primary pipe rupture (Category 4)
- Inadvertent withdrawal of control rod (Category 2)

- Sudden acceleration of one or more primary pumps (Category 2)
- Sudden acceleration of one or more secondary pumps (Category 2)
- Loss of feedwater flow/boiler feed pump trip (Category 2)
- Sudden increase of water flow (Category 2)

1.8 PLANT PROTECTION SYSTEM

The plant protection system (PPS) is to ensure that specified safety limits (SL) on operating parameters are not reached for all postulated design basis events. It should provide the required protection by sensing the need for and carrying to completion reactor trip, pump trip, turbine trip, etc. The concept of safety limits is based on the prevention of unacceptable releases of radioactive materials from the plant through the application of limits imposed on the temperatures of fuel and fuel cladding, coolant pressure, pressure boundary integrity, and other operational characteristics influencing the release of radioactive material from the fuel. Established safety limits are to protect the integrity of certain physical barriers that guard against the uncontrolled release of radioactive material. The safety limits should be established by means of a conservative approach to ensure that all the uncertainties of safety analyses are considered.

Limiting safety system settings (LSSS) are the parameters included in safety limits as well as other parameters, or combinations of parameters, which could contribute to pressure or temperature transients (IAEA 2000). Exceeding some such settings will cause the reactor to be tripped to suppress a transient. Exceeding other settings will result in other automatic actions to prevent safety limits from being exceeded. Thus, the safety limit on power would be that power beyond which operation is unsafe, while the LSSS would be the one at which trip is initiated. LSSS should consider measurement and instrumentation uncertainties associated with the process variable.

Limiting conditions for operation (LCO) are intended to ensure safe operation—that is, to ensure that the assumptions of the safety analysis report are valid and that established safety limits are not exceeded in the operation of the plant. In addition, acceptable margins should be ensured between the normal operating values and the established safety system settings to avoid undesirably frequent actuation of safety systems. Figure 1.6 demonstrates a correlation between safety limits, safety system settings, and limits for normal operation.

In Figure 1.6, the range of steady-state operation refers to the monitored parameter being kept within the steady-state range by the control system or by the operator in accordance with the operating procedures. The monitored parameters may exceed the steady-state range because of load changes or imbalance of the control system, for example, curve 1. If the temperature reaches an alarm setting, the operator will be alerted and will act to

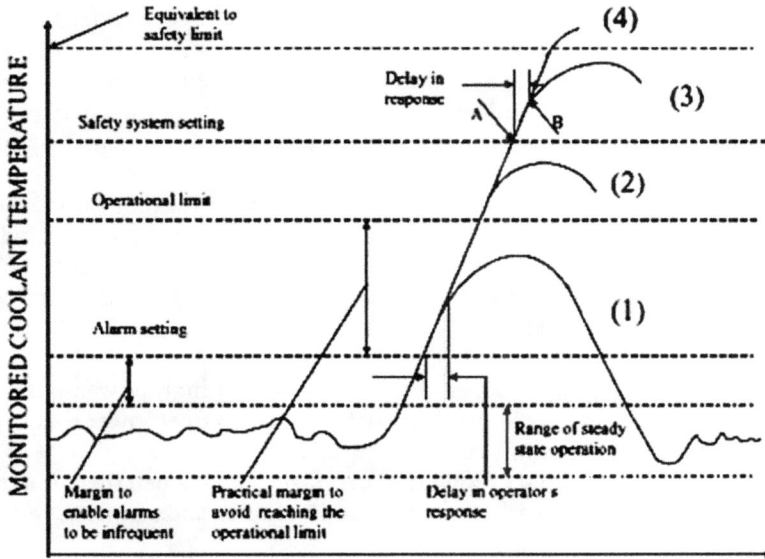

Figure 1.6 Interrelationship between a safety limit, safety system setting, and operational limit.

Source: IAEA Operational limits and conditions and operating procedures for nuclear power plants: safety guide, 2000.

supplement any automatic systems in reducing temperature to the steady-state values without allowing the temperature to reach the limit for normal operation. The delay in the operator's response should be taken into consideration.

Limits for normal operation may be set at any level between the range of steady-state operation and the actuation setting for the safety system, based on the results of the safety analysis. It is normal to have margins between alarm settings and operational limits to take account of routine fluctuations arising in normal operation. There may also be a margin between the operational limit and the safety system setting to allow the operator to take action to control a transient without activating the safety system. If the operational limit is reached and the operator can take corrective action to prevent the safety system setting being reached, then the transient will be of the form of curve 2.

In the event of malfunction of the control system or operator error or for other reasons, the monitored parameter might reach the safety system setting at point A (curve 3) with the consequence that the safety system being actuated. This corrective action only becomes effective at point B owing to inherent delays in the instrumentation and equipment of the safety system. The response should be sufficient to prevent the safety limit being reached, although local fuel damage cannot be excluded.

In the event of a failure that exceeds the most severe one that the plant was designed to cope with, or a failure or multiple failures in a safety system, it would be possible for the temperature of the cladding to exceed the value of the safety limit (curve 4), and hence significant amounts of radioactive material could be released. Additional safety systems may be actuated by other parameters to bring other engineered safety features into action to mitigate the consequences, and measures for accident management may be activated.

Plant protection system (PPS) can be taken as a control system, which remains as an observer during normal operation and acts when the plant system reaches the limit of permissible operation. The following are common protective parameters:

1. High Neutron Flux: Generates a trip signal on high power, based on the fuel temperature limits. Mostly for transient overpower events due to reactivity insertion.
2. Neutron Flux/Primary Flow: Initiates a trip signal when there is a transient overpower or a large reduction in reactor sodium flow, referred to as an undercooling event.
3. Reactor Flow: Initiates a trip for any undercooling event due to pump trip or seizure.
4. Core Outlet Temperature: Initiates a trip action based on processing of the sodium outlet temperatures of the different subassemblies. This responds to both undercooling and overpower events.
5. Delayed Neutron Detection (DND): Initiates a trip when the delayed neutron detection in sodium crosses the threshold. It is measured by sampling the sodium coming from the core and passing the same over a neutron detector, which will indicate presence of fission products pointing out to a fuel failure. The idea of a trip is to prevent further fuel failure and contamination of reactor sodium.
6. Loss of Off-site Power: Initiates a trip as the primary cooling is lost.

Incidents in the secondary and steam water system affect the heat removal function and result in the increase of reactor inlet temperature, and this can be used to trip the reactor. However, from the point of having diverse signals to ensure a reliable reactor trip, other trip signals that originate from secondary sodium system are a secondary sodium pump trip, low level of sodium in capacities due to say sodium leak, etc. From the steam water system, a boiler feed pump trip and turbine trip signals would initiate a reactor trip.

1.9 SENSORS AND RESPONSE TIME

For finalizing the LSSS, the response time of the different sensors is to be considered. Thermocouples at core outlet have a typical response time of $4 \pm 2s$ (Value for FBTR) (Vaidyanathan et al., 1987). These values are to be based on the tests carried out on the thermocouple-thermowell combination. A response time of 6 seconds is considered in the simulation. The sodium flow measurement using electromagnetic flow meters and related electronics has a response time of 1 second. Neutron flux measurements using boron counters in the low power range, fission counters in the intermediate power range, and compensated ion chambers in the power range up to full load have response times < 1 second. For the signals from the electrical contacts of the pumps in case of loss of power, though signal is instantaneous, the trip is initiated with a delay of ~3 seconds to avoid trips for short-duration power failures. There is a need to minimize the number of trips, as each shutdown gives a cold shock to the components. Also, the signals are triplicated with a 2/3 logic, with two out of three being sufficient to initiate the trip or safety system action. The requirement that two channels must both vote for a trip reduces the likelihood of spurious trips due to a single component failure. On the other hand, the reactor will still trip if required even if one channel is unavailable (failed unsafe). Finally, a single channel can be tested without tripping the reactor. Such a scheme allows testing of each channel without causing a trip, thereby improving availability. There is also the need to consider the response time of the shutdown system, which is ~1 second for FBTR in finalizing the LSSS.

1.10 SCOPE OF DYNAMIC STUDIES

One of the important inputs for design is the coast-down law of the primary sodium pump. During power failure to the primary pumps, the primary flow should not reduce faster than the power, which is reduced by action on control rods. Gradual coast-down is achieved by having a flywheel on the drive system. Too large a flywheel may complicate the drive system, and too small flywheel will lead to faster flow reduction resulting in larger thermal shocks to sodium-immersed structures. Optimization is done by analyzing the core thermal and hydraulic models for various coast-down rates (different flywheel inertias) and their impact on plant parameters. The dynamic models must be utilized to identify the parameters that could be used for automatic safety actions for the different events at the plant. Some of the direct parameters are power, outlet temperature of sodium from different subassemblies, pump flow, etc. Parameters like reactivity and power/flow are derived quantities that provide diversity in safety parameters. The models also are effectively used to assess the influence of uncertainties in the various property data of fuel and structures used. These, then, are useful to specify the tolerances for fuel fabrication. The parallel operation of primary pumps across the reactor core imposes a minimum speed difference, as otherwise there could be reverse flow through the pump running at lower speed. The elevation of intermediate heat exchanger (IHX) with respect to the core also plays an important role in assuring natural circulation in the primary sodium system during power failure conditions. All the aforementioned aspects demand a dynamic simulation of thermal hydraulic processes in the plant.

The secondary sodium circuit design inputs comprise the coast-down law for the secondary sodium pump, the sizing of the surge tank, etc. The IHX thermal model analysis leads to the acceptable secondary coast-down law for a given primary sodium coast-down, to limit the thermal shock in IHX to acceptable values based on stress evaluation. The surge tank sizing is based on its capacity to limit the pressures seen by the IHX in case of a sodium water reaction, in the steam generator, which causes a pressure surge in the secondary sodium. The elevation of a steam generator with respect to IHX has a role in promoting natural convection if the steam generator is in the decay heat removal path.

The above-mentioned objectives can be realized by resorting to dynamic simulation of the plant components. The individual models can be coupled into an integrated plant dynamics model, which could be used effectively to study the combination of perturbations. Much critical data used in simulation need to be found out experimentally due to a complex nature of the geometry or process, e.g., pressure drop coefficient of fuel subassembly, hot pool hydraulics, etc.

1.11 MODELING DEVELOPMENT

Models and associated computer codes have been developed in different countries for their FBR programs. Computer codes IANUS (Additon et al., 1976) and DEMO (Alliston, 1978) were developed for the fast flux test facility (FFTF) reactor and the clinch river breeder reactor plant (CRBRP) in the USA to simulate the overall response of a loop-type plant. FFTF plant is a loop-type FBR with the final sink as sodium to air dump heat exchanger, while CRBRP, also of loop type, has the steam generator and the associated steam water system. IANUS contains the models of the reactor, the primary and secondary heat transports, and the dump heat exchanger. In DEMO, the dump heat exchanger is replaced with the model for the steam generator. Another code—EPRI-CURL, like DEMO but faster running—was developed at Cornell University (Khatib and Cady, 1981). The code NALAP (Martin et al., 1975) was developed using the base of RELAP3 code used for light-water reactors, by substitution of the water properties with that of sodium. NATDEMO was developed for a pool-type EBR-II reactor (Mohr and Feldman, 1981). NATRANS was another code developed in Germany for the loop-type SNR 300 reactor (Broache, 1971). In the UK, for analyzing the transient in a protype fast reactor (PFR), a computer code MELANI was created (Durham, 1976). A general-purpose plant dynamics code SSC-L was developed for loop-type fast reactors in the Brookehaven National Lab, USA (Agrawal and Guppy, 1978). This code provides detailed models of the core, including sodium boiling. This code was developed as part of USNRC's licensing studies. Another code, SSC-K, was developed in Korea for their pool-type KALIMER reactor (Won-Pyo Chang et al., 2002). In France, OASIS code (Dufour, 1995) and, more recently, CATHARE code (Tenchine et al., 2012) were developed for sodium-cooled fast reactors. Each code has several simplifying assumptions and approximations.

Recently, under the coordinated research program of IAEA, the natural convection tests conducted in EBR II and the French PHENIX reactor have been used to validate many computer codes used in the dynamic simulation of sodium-cooled fast reactors. Details of some of the codes that participated in the benchmark exercise are given in Appendix A. It gives useful insight into the modeling capabilities of recent computer codes.

ASSIGNMENT

1. Indicate the differences in fission in a PWR and a BWR.
2. Explain in brief the basics of breeding in PWRs and SFRs.
3. Compute ETA (η) values for different fissile materials based on data available in literature and compare with those indicated in Figure 1.1.
4. "Effective utilization of uranium resources is possible only in fast reactors." Is this statement true? If true, list the facts in support of the statement. If false, indicate the same with supporting statements.
5. What is the need for imposing a secondary sodium system between the reactor and steam water circuit in an SFR unlike a PWR?

REFERENCES

Additon S.L., McCall T.B. and Wolfe, C.F. (1976), "Simulation of the overall FFTF plant performance", *Hanford Engg. Dev. lab*, HEDL-TC-556.

Agrawal A.K. and Guppy J. (1978), "An advanced thermohydraulic simulation code for transients in LMFBRs", Brookhaven national Laboratory, BNL-NUREG-50773.

Agrawal A.K. (1983), "Comparison of CRBR design basis events with those of foreign LMFBR plants", NUREG/CR-3240/BNL/NUREG-51663, USNRC.

Alliston W.H. (1978), "LMFBR demonstration plant simulation model-DEMO", *Westinghouse Adv. Reactor. Div.*, CRBRP-ARD-0005.

Bahman Z. (2020), 6 - Generation IV nuclear reactors, Editor(s): Salah Ud-Din Khan, Alexander Nakhabov, In Woodhead Publishing Series in Energy, *Nuclear Reactor Technology Development and Utilization*, Woodhead Publishing, pp. 213–246.

Broache D. (1971), "Report MRP-71", Garching, Germany.

Dufour P. (1995), "OASIS: An interactive simulation case", Simulation Multiconference, Phoenix, Arizona, April 1995.

Durham M.T. (1976), "Influence of reactor design on the establishment of natural circulation in pool type LMFBR", *J. Brit. Nucl. Energy*, Vol. 305, 305–310.

ERRI NP-359, (1977), Assessment of Thorium Fuel Cycles in Pressurized Water Reactors, Final Report, February 1977.

IAEA, (2013), *Design features and operating experience of experimental fast reactors*, International Atomic Energy Agency, Vienna.

IAEA, (1996), Defence in depth in nuclear safety: INSAG-10, A report by the International Nuclear Safety Advisory group, International Atomic Energy Agency, STI/PUB/1013.

IAEA Safety Standards Series No. NS-G-2.2, 2000, Operational Limits and Conditions and Operating Procedures for Nuclear Power Plants, International Atomic Energy Agency, Vienna.

IAEA Safety Standards Series No. SSG-2 (Rev. 1), 2019, Deterministic Safety Analysis for Nuclear Power Plants Specific Safety Guide.

Kasinathan N., Parthasarathy U., Natesan K., Selvaraj P., Chellapandi P., Chetal, S.C. and Bhoje S.B. (2002), "Analysis of design basis events for prototype fast breeder reactor", (Editor.): Gupta, Satish K. In *Nuclear reactor safety*, First BRNS conference on Nuclear Reactor Technology, Bhabha Atomic Research Centre, Mumbai.

Khatib Rahbar M. and Cady K.B. (1981), "Dynamic models and associated numerical simulation of system wide transients in loop type LMFBR", *Nuc. Eng. Des*, Vol. 64, p. 259.

Martin B.A., Agrawal A.K., Albright D.C., Epel L.G. and Maise G. (1975), "NALAP: An LMFBR system transient code", BNL-50457.

Mohr D. and Feldman E.E. (1981). "A dynamic behavior of the EBR-II plant during natural convection with NATDEMO code", Editor(s): Agrawal, A.K., Guppy J.G., In *Decay heat removal and natural convection in FBRs*, Hemisphere Pub., New York.

Raj B., Chellapandi, P. and Rao, P. V. (2015), *Sodium Fast Reactors with Closed Fuel Cycle*, CRC Press, doi:10.1201/b18350

Tenchine D., Bariere R., Bazin P., Ducros F., Kadri D., Rameau B. and Tauveron N. (2012), "Status of CATHARE code for sodium cooled fast reactors", *Nucl. Eng. and Des*, Vol. 245, p. 140; doi:10.1016/j.nucengdes.2012.01.019.

Vaidyanathan G., Sangodkar D.B. and Paranjpe S.R. (1987), Limiting safety system settings for FBTR operation, Load following control of nuclear power plants including availability aspects, Proceedings of the specialists' meeting, Bombay during 10–12 Dec 1985.

Wade D. C. and Hill R. N. (1997), The design rationale of the IFR, *Prog. Nucl. Energy*, Vol. 31, No. 1–2, pp. 13–42: doi:10.1016/0149-1970(96)00002-9.

Waltar A.E., Donald R. Todd, Pavel V. (2012), Tsvetkov, Editors, *Fast Spectrum Reactors*, Springer, doi:10.1007/978-1-4419-9572-8.

Won-Pyo Chang, Young-Min Kowan, Yong-Bum Lee and Dohee Hahn (2002), "Model Development for analysis of the Korea advanced liquid metal reactor", *Nucl. Eng. Des*, Vol. 217, p. 63; doi:10.1016/S0029-5493(02)00129-2.

Chapter 2

Description of Fast Reactors

2.1 INTRODUCTION

Many experimental and prototype fast reactors cooled by sodium have been built around the world and accumulated more than 400 years of experience. As a precursor to dynamic simulation, it is necessary to have knowledge of the different systems, components, and their interdependency to ensure safe operation. As a first step to simulation, one needs to know the details of the various systems and components and their role in ensuring the safety of the plant. In this chapter a brief description of the loop-type experimental fast breeder test reactor (FBTR) and pool-type prototype fast breeder reactor (PFBR) from India is offered. FBTR is in operation since 1985, while PFBR is in the commissioning stage.

2.2 FAST BREEDER TEST REACTOR (FBTR)

Fast breeder reactors constitute the second stage of India's three-stage nuclear energy program for effective utilization of the country's limited reserves of natural uranium and exploitation of its large reserves of thorium. FBTR was built with transfer of technology based on the French reactor RAPSODIE (Bhoje et al., 1985). FBTR has several design changes vis-a-vis Rapsodie. Major changes include the use of Pu–U monocarbide fuel in place of enriched uranium oxide and incorporation of steam generators and a turbine generator in place of a sodium of air terminal heat exchanger.

FBTR is a 40MWt, loop-type, sodium-cooled fast reactor. It reached full power in March 2022. Steam conditions are 190/480°C, 125 Kg/cm^2. Figure 2.1 shows the schematic flow sheet of the heat transport circuits (Srinivasan et al., 2006). Heat generated in the reactor is removed by two primary sodium loops and transferred to the corresponding secondary sodium loops. Each secondary sodium loop is provided with two steam

DOI: 10.1201/9781003283188-2

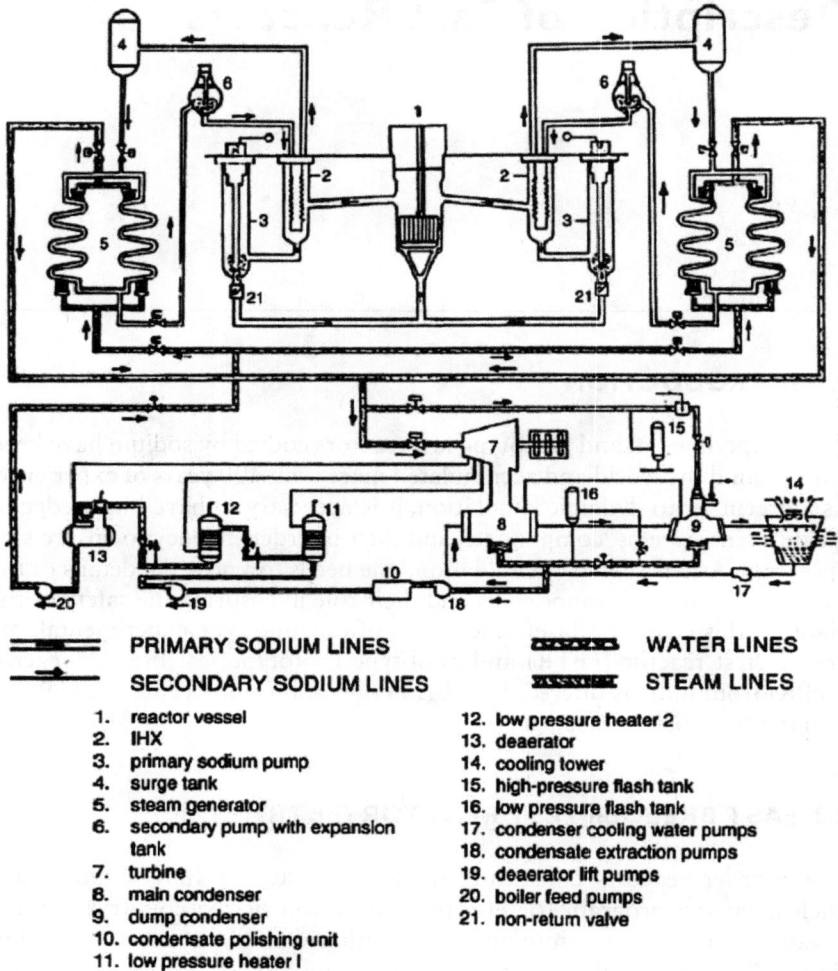

PRIMARY SODIUM LINES		**WATER LINES**	
SECONDARY SODIUM LINES		**STEAM LINES**	

1. reactor vessel	12. low pressure heater 2
2. IHX	13. deaerator
3. primary sodium pump	14. cooling tower
4. surge tank	15. high-pressure flash tank
5. steam generator	16. low pressure flash tank
6. secondary pump with expansion	17. condenser cooling water pumps
tank	18. condensate extraction pumps
7. turbine	19. deaerator lift pumps
8. main condenser	20. boiler feed pumps
9. dump condenser	21. non-return valve
10. condensate polishing unit	
11. low pressure heater I	

Figure 2.1 FBTR heat transport system − schematic.

Source: Srinivasan et al., *The Fast Breeder Test Reactor—Design and operating experiences Nuclear Engineering and Design* 236, 2006.

generator modules. Steam from the four modules is fed to a common steam–water circuit comprising a turbine generator. Stainless steel (SS 316) is the principal material of construction for the reactor and the coolant circuits.

Since the turbine generator was a prototype design, it was needed that the unavailability of the turbine should not affect the reactor power operation. Hence a 100% dump stem system was provided across the turbine. It also helps maintain a reactor operation in case of a spurious turbine trip.

2.2.1 Reactor Core

The core (Figure 2.2) consists of 745 closely packed locations, with fuel at the center, surrounded by nickel reflectors, thoria blankets, and steel reflectors. The fuel is 70% PuC + 30% UC. The core is vertical and freestanding, with the subassemblies supported at the bottom by the grid plate and attached to the latter by collapsible hold-down springs. The subassemblies are hexagonally shaped. The core has provision for 65 fuel subassemblies, 3 test steel subassemblies, 6 control rods, 143 nickel reflectors, 342 thoria blankets, and 163 steel reflector subassemblies. In addition, there are 23 storage locations in the outermost steel reflector row. The fuel subassembly (Figure 2.3) is 1661.5 mm long, with a width across flats of 49.8 mm, and houses 61 fuel pins. The fuel pins are 531.5 mm long, with fuel stack length

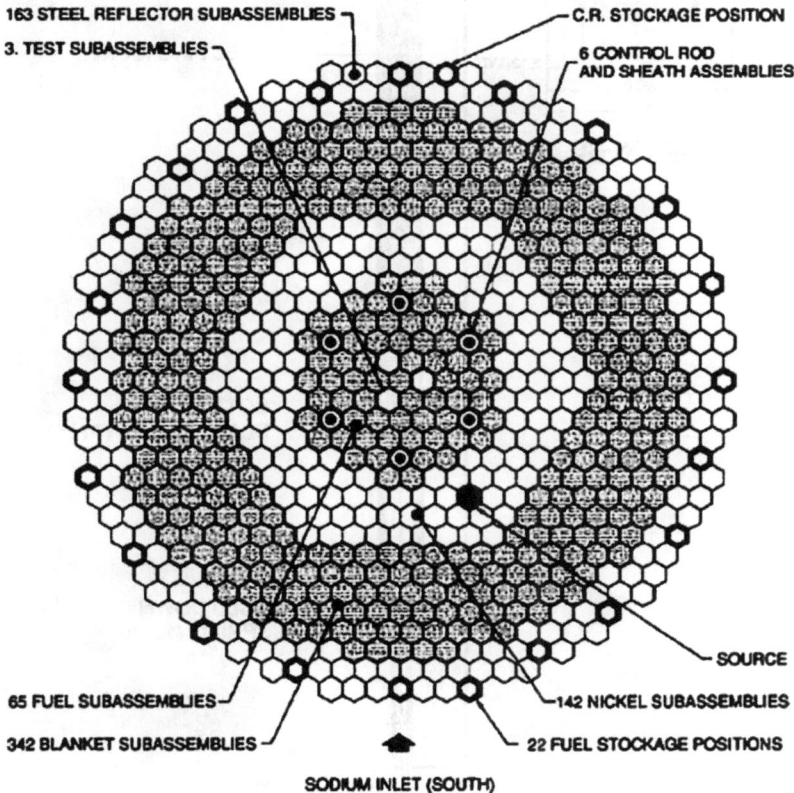

Figure 2.2 FBTR reactor core plan.

Source: Srinivasan et al., The Fast Breeder Test Reactor—Design and Operating Experiences. *Nuclear Engineering and Design* 236, 2006.

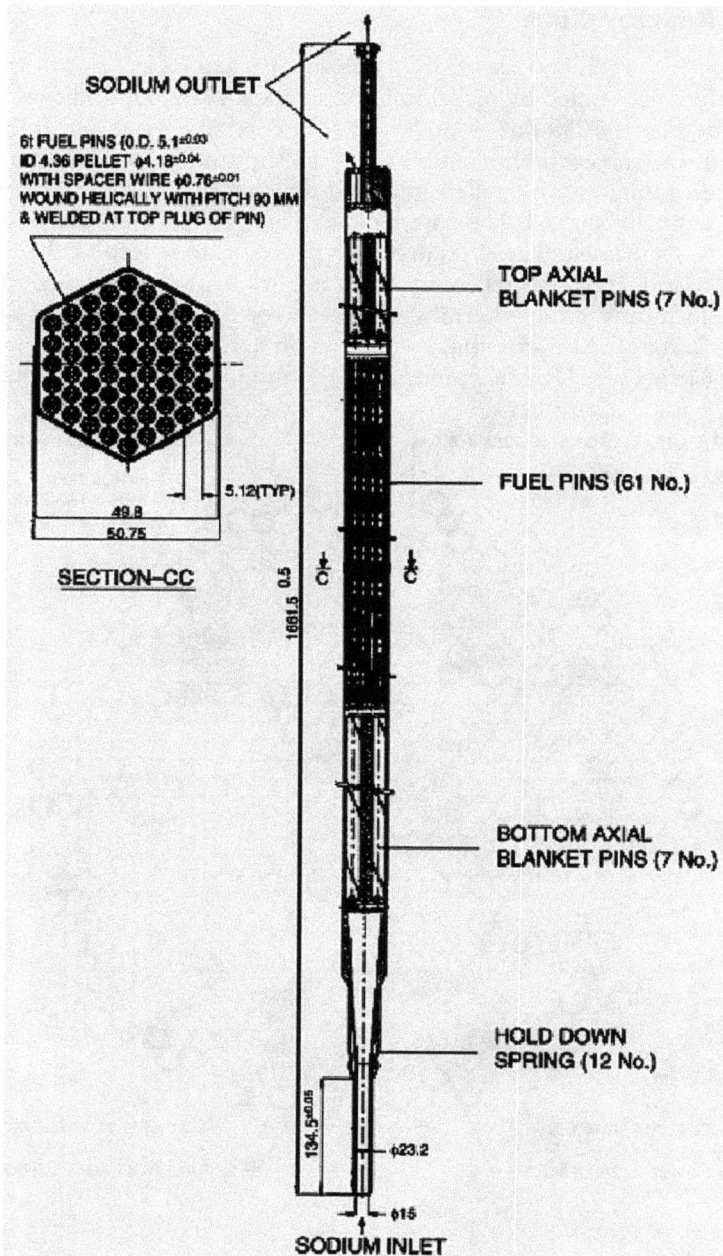

Figure 2.3 Fuel subassembly.

Source: Srinivasan et al., The Fast Breeder Test Reactor—Design and Operating Experiences. *Nuclear Engineering and Design* 236, 2006.

of 320 mm. Axial blankets of 26 3mm in length are provided on either side of the fuel pins.

2.2.2 Reactor Assembly

The reactor vessel houses the core and serves as a conduit for the primary sodium coolant flow through the core (Figure 2.4). Charging and discharging

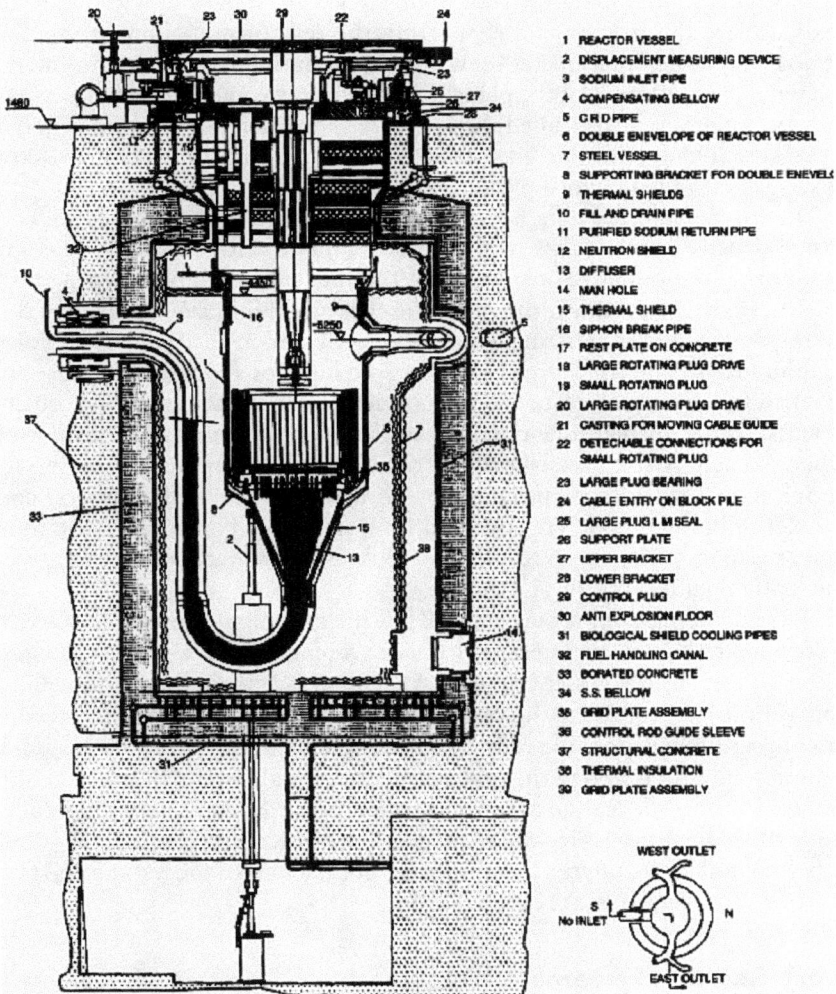

1 REACTOR VESSEL
2 DISPLACEMENT MEASURING DEVICE
3 SODIUM INLET PIPE
4 COMPENSATING BELLOW
5 GRID PIPE
6 DOUBLE ENEVELOPE OF REACTOR VESSEL
7 STEEL VESSEL
8 SUPPORTING BRACKET FOR DOUBLE ENEVELO
9 THERMAL SHIELDS
10 FILL AND DRAIN PIPE
11 PURIFIED SODIUM RETURN PIPE
12 NEUTRON SHIELD
13 DIFFUSER
14 MAN HOLE
15 THERMAL SHIELD
16 SIPHON BREAK PIPE
17 REST PLATE ON CONCRETE
18 LARGE ROTATING PLUG DRIVE
19 SMALL ROTATING PLUG
20 LARGE ROTATING PLUG DRIVE
21 CASTING FOR MOVING CABLE GUIDE
22 DETECHABLE CONNECTIONS FOR SMALL ROTATING PLUG
23 LARGE PLUG BEARING
24 CABLE ENTRY ON BLOCK PILE
25 LARGE PLUG L M SEAL
26 SUPPORT PLATE
27 UPPER BRACKET
28 LOWER BRACKET
29 CONTROL PLUG
30 ANTI EXPLOSION FLOOR
31 BIOLOGICAL SHIELD COOLING PIPES
32 FUEL HANDLING CANAL
33 BORATED CONCRETE
34 S.S. BELLOW
35 GRID PLATE ASSEMBLY
36 CONTROL ROD GUIDE SLEEVE
37 STRUCTURAL CONCRETE
38 THERMAL INSULATION
39 GRID PLATE ASSEMBLY

WEST OUTLET

No INLET

EAST OUTLET

Figure 2.4 FBTR reactor assembly.

Source: Srinivasan et al., The Fast Breeder Test Reactor—Design and Operating Experiences. *Nuclear Engineering and Design* 236, 2006.

of subassemblies are done from the reactor top with the reactor in a shut-down state. The sodium inlet pipe joins the reactor vessel at the bottom, and two sodium outlet pipes radially branch out of the vessel above the core. The entire primary sodium circuit is provided with a nitrogen-filled enve-lope called a *double envelope*, designed to minimize the sodium level drop in the reactor in the event of any sodium leak. The reactor is closed at the top by large and small rotatable plugs, which serve as top shields and enable access to the core locations for fuel handling. The rotatable plugs are cooled by nitrogen. Liquid metal seals of tin–bismuth alloy, backed with inflatable seals, isolate the reactor cover gas from the reactor containment building atmosphere. The liquid metal seals are frozen during reactor operation and melted during rotation of the plugs. The space between the liquid metal seals and the inflatable seals, called the *interseal space*, is maintained in argon at a pressure higher than the reactor cover gas to prevent the leakage of active cover gas into the reactor building.

The small rotatable plug houses the control plug, which carries six con-trol rod drive mechanisms and core cover plate with thermocouples for monitoring the outlet temperatures of the fuel subassemblies. Ten neutron shields, each 25 mm thick, surround the core and minimize the incident flux on the reactor vessel. Thermal shields are provided inside the reactor vessel to minimize the thermal stresses due to cold and hot shocks. The radial and axial shifts of the grid plate are monitored by two displacement-measuring devices. Large displacement may result in non-insertion of control rods when needed. A steel vessel with thermal insulation surrounds the reactor vessel. Radial shielding is provided by borated concrete and structural con-crete. The borated concrete is cooled by water pipes embedded close to its inner periphery. The entire reactor assembly is suspended from the top, with the load taken by structural concrete.

During commissioning stages in FBTR, the grid plate supporting the core was found to be shifting laterally (by about 8 mm) due to temperature asym-metry in the reactor vessel (Figure 2.5). This was essentially due to cellular convection of hot argon in the vertical annuli in the top closures. Suppression of convection was possible only by injecting helium into the annulus while maintaining argon above the sodium. The thermal conductivity of helium being ~10 times more than argon, the temperature differentials were smaller. With this the tilting issue was managed (Suresh Kumar et al., 2011). Both experimental and analytical studies were conducted to understand the phe-nomena (Hemanath et al., 2007).

2.2.3 Sodium Systems

Primary sodium is pumped into the reactor by primary sodium pumps and flows by gravity to the intermediate heat exchangers and then back to the pump suction. The intermediate heat exchangers are vertical, counter-flow heat exchangers that transfer heat from the active primary sodium to the

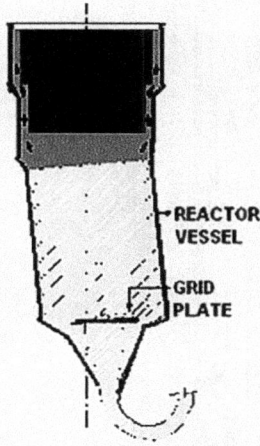

Figure 2.5 Reactor vessel tilt due to argon convection.

Source: Suresh Kumar et al., Twenty-Five Years of Operating Experience with the
    Fast Breeder Test Reactor, *Energy Procedia* 7, 2011.

inactive secondary sodium. Primary sodium flows on the shell side and secondary sodium on the tube side. The secondary sodium is at a higher pressure than primary sodium in the IHX (Figure 2.6). This is essentially to ensure that radioactive primary sodium does not enter into secondary sodium in case of a tube leak. The shell is fixed, and the tube bundles are removable. Secondary sodium is pumped into the intermediate heat exchangers by secondary sodium pumps. After removing heat from primary sodium, the secondary sodium enters the steam generators. A surge tank with free level and topped up by cover gas argon is interposed between the intermediate heat exchangers outlet and steam generator inlet. This tank acts as a buffer against pressure wave transmission to intermediate heat exchangers during sodium–water reaction in steam generators due to any water leaks. The sodium after exchanging heat in the steam generator passes through an expansion tank in which the secondary pump is also located before entering the IHX. The sodium pumps are vertical, single-stage centrifugal pumps with axial suction and radial discharge. Each pump has a fixed shell and a removable assembly comprising the impeller, the diffuser, and the shaft. The shaft is supported by taper roller bearings at the top and guided by hydrostatic bearings at the bottom.

For safety reasons, there are no valves in the primary and secondary sodium main loops. Flow control is by controlling the speeds of the pumps. The pumps are driven by direct current motors and powered by a motor generator set. Flywheels mounted on drives provide sufficient inertia to run the pumps in the event of power failure, to ensure that fuel clad hot spot temperature is maintained within limits. The pump drives are provided with

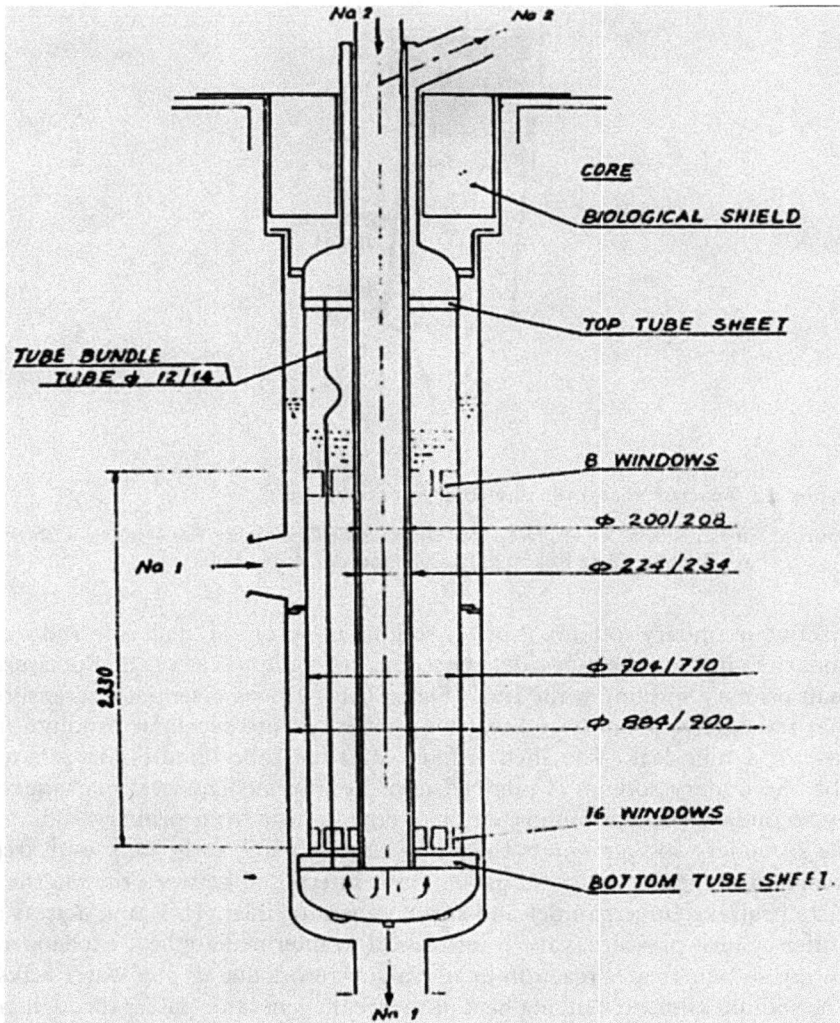

Figure 2.6 IHX in FBTR.

emergency power supply from the station diesel generators, and battery backup is provided for the primary pump drives to provide adequate flow for safe removal of decay heat.

All the sodium capacities are provided with an inert cover of argon above the free sodium levels. Argon purity is maintained through sodium-potassium alloy (NaK) bubblers that will absorb the oxygen impurity in argon. Both primary and secondary sodium systems are provided with cold traps for sodium purification. These employ the principle of precipitation of oxide

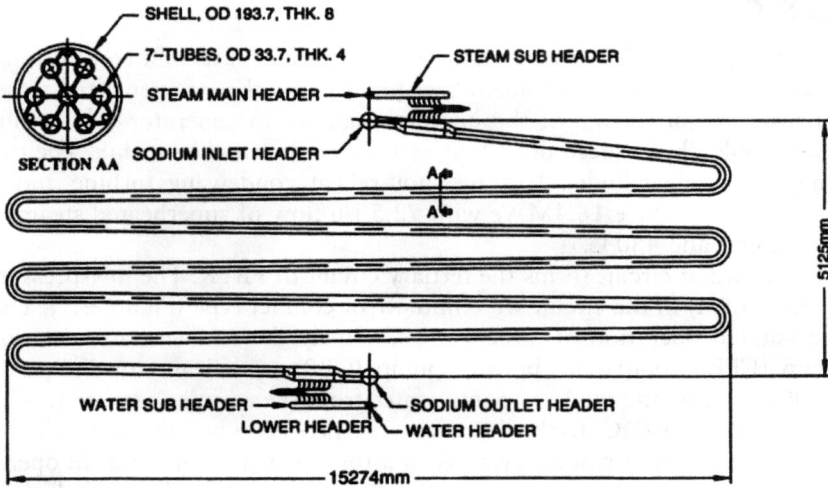

Figure 2.7 FBTR steam generator.

Source: Srinivasan et al., The Fast Breeder Test Reactor—Design and Operating
  Experiences. *Nuclear Engineering and Design* 236, 2006.

impurities when the sodium temperature is brought down below the saturation temperature values. Plugging indicators monitor sodium purity and
work on a similar principle.

The steam generator modules are of once-through, counterflow type, with
sodium entering the shell side from top and water entering the tube side
from bottom (Figure 2.7). The modules have a serpentine configuration,
with evaporation and superheating occurring in a single pass. Such a configuration helps in absorbing the differential expansion between the tube
bundle and shell. Due to the absence of a drum and provisions for blowdown, feed water chemistry is maintained within very stringent limits using
100% demineralizer units.

2.2.4 Decay Heat Removal

With availability of grid power, the decay heat is removed through the normal steam water circuit. The steam generator modules are housed inside an
insulated casing. By opening the trap doors of the casing, it is possible to
remove decay heat from the reactor by natural convection, when the steam
water circuit is not available. The annular gap between the reactor vessel
and its double envelope is used for emergency cooling of the core during the
very-low-probability incident of simultaneous rupture of coolant boundary
and its double envelope outside the reactor.

2.2.5 Generating Plant

The steam–water circuit consists of all the equipment as in a conventional power plant. An online demineralizer plant meets the stringent feed water chemistry requirements of the once-through steam generators. A cooling tower cooled by induced draft fans serves as a terminal heat sink. The turbine is a single-cylinder, 16-stage, non-reheat condensing turbine and is designed to produce 16.4MWe with 72.5 t/h flow of superheated steam at 120 kg/cm^2 and 480°C.

Steam water circuit forms the tertiary circuit of FBTR. The low-pressure heaters (LPH) in the circuit were initially of contact type (Figure 2.1). This necessitated operation of three pumps, namely the condensate extraction pump (CEP), condensate booster pump (CBP), and deaerator lift pump (DLP), in a cascade mode to pump water from the main condenser (MC) / dump condenser (DC) to the deaerator. The provision of a dump condenser allows operation of reactor even when a turbine generator is not in operation. Due to cascade operation of three pumps in series in a condensate system, which takes suction from capacities operating with a free water level, there were numerous reactor trips due to fluctuations in levels in the LP heaters even during small process / grid disturbances. To overcome this problem, the contact-type LP heaters were replaced by conventional shell and tube surface type heaters in 2004. The condensate booster pump between LPH-1 and LPH-2 was also dispensed with. After this modification, there was no incident of a reactor trip from a steam water system.

2.2.6 Instrumentation and Control

The reactor power control and shutdown are accomplished by six control rods. For shutdown, the rods are inserted into the core in two modes: lowering of rods, wherein all the rods are driven down by the respective drive mechanisms at a speed of 1 mm/s; and SCRAM, wherein the rods drop down by gravity in less than 400 ms. SCRAM is ordered by neutronic parameters (power, period, and reactivity), core thermal parameters, and delayed neutron detection system. Lowering of rods is ordered by thermal hydraulic parameters of the plant, which, because of the thermal inertia, do not call for fast shutdown. Power control is manual.

Reactivity is not a direct signal but dynamically computed from the neutron flux and period. Most of these digital reactivity meters employ reactivity estimation based on inverse point kinetic equations. Kalman filtering algorithm, with its inherent ability to work with even highly noisy input signals, is used to compute the reactivity (Patil and Shimjith, 2014).

A central data processing system comprising two computers (one online and the other on auto-standby) monitors the core outlet temperatures and generates a reactor trip on mean core outlet temperature, mean core temperature rise, and plugging of any fuel subassembly. The central data processing system also initiates the lowering of rods on control rod level discordance and rate of rise of the steam generator leak detection system signal. In addition, it checks the healthiness of the reactor protection system by carrying out a fine impulse test (FIT) and supervises the analog and digital signals. Sodium levels in the capacities are monitored by continuous and discontinuous level probes. Flows are monitored by permanent magnet-type electromagnetic flow meters. Sodium leak is detected by spark plugs, wire type, and ionization types of detectors.

A water leak into sodium in the steam generator at the incipient stage (50 mg to 10 g/s) is detected by a very sensitive mass spectrometer–based steam generator leak detection system to measure hydrogen in sodium in ppb levels. At lower sodium temperatures, as during valving in water into the steam generators during power raising, a leak is detected by monitoring the hydrogen in the cover gas of expansion tanks. Medium leaks (0.01–0.5 kg/s) are detected by monitoring the expansion tank cover gas pressure. Quick-closing valves isolate the steam generators on the sodium and water sides in the event of large leaks (>0.5 kg/s), and rupture discs provided in the sodium headers relieve pressure build-up.

2.2.7 Safety

The reactor is protected against transient overpower accident by feedback through negative temperature and power coefficients. Interlocks are provided to ensure that only one control rod can be operated at a time. Trip is provided on discordance of control rod levels in the event of any uncontrolled withdrawal. The reactor is provided with redundant and diverse trips from neutronic and core thermal parameters, i.e., Lin-P, log-P, period, reactivity, mean core outlet temperature, and mean core temperature rise. Insertion of two control rods is sufficient to shut down the reactor from the most reactive configuration.

The double envelope of the primary sodium loops safeguards against a loss-of-coolant accident. Leak-before-break enables early detection before any leak escalation. Spark plugs at the bottom of the primary sodium capacities and ^{24}Na particulate activity monitors in the nitrogen in the double envelope enable detection of a leak at the incipient stage. Level probes in capacities confirm escalated leaks. In the event of any leak in a primary circuit, the sodium level in the reactor stabilizes above the outlet pipes, and circulation by pumps is maintained. Engineered safety is also provided for the very improbable incident of failure of the main coolant boundary and the double envelope. If the leak is outside the reactor, complete draining of the reactor is avoided by the siphon break pipe, which communicates between the reactor inlet pipe and sodium in the reactor. The sodium level in the reactor in this incident stabilizes above the core but below the outlet pipe. Under these stagnant conditions, the reactor is cooled by circulating nitrogen in the annular space between the reactor vessel and its double envelope. If failure of the main coolant boundary and the double envelope occurs inside the reactor assembly, the leaked sodium collects in the steel vessel. There is a provision to flood the reactor vessel with sufficient sodium from a dedicated sodium flooding system to ensure submergence of the core. Biological shield cooling system removes heat from the steel vessel.

Absence of any valve in the loops and operation of two pumps in parallel safeguard against a loss-of-flow accident. Trips are provided from total flow through the reactor, individual loop flow, individual pump trip, core outlet temperature, and core temperature rise. During any off-site power failure, the reactor trips and primary pumps coast down due to the inertia of the flywheels, with a flow-halving time of 17 s. The pumps meanwhile are taken over by the emergency diesel power. If a station blackout occurs, dedicated battery banks take over and run the pumps at 150 rpm for at least 30 min. Decay heat removal beyond this time is possible by natural convection by opening the trap doors of the steam generator casing.

Subassemblies are flow tested in nitrogen before loading in the reactor. Plugging of any subassembly in the reactor is avoided by providing radial entry sleeves in the grid plate. Plugging during operation is detected by a plugging detection subroutine in the central data processing system, which initiates alarm, lowering of rods, and SCRAM when any fuel subassembly outlet temperature exceeds its set limits. Void coefficient of reactivity being negative at all positions of the core, sodium boiling due to flow starvation of any subassembly adds negative reactivity. In the event of clad failure, cover gas activity alarm sounds, followed by trip from the delayed neutron detection system.

The scenarios studied for design basis accidents are: transient overpower accident (due to uncontrolled withdrawal of control rod), loss-of-coolant accident, and loss-of-flow accident, without protection, i.e., without the actuation of the shutdown system. The resulting scenario under any of these conditions is partial core melting, rapid rise of power due to additional reactivity caused by slumping of the core, fuel vaporization, and core disassembly. Theoretical calculations indicate the mechanical energy release due to instantaneous isentropic expansion of the fuel vapor as 14.5 MJ or less than 3.5 kg of TNT. Reactor vessel can safely contain 8.8 kg of TNT. The energy released through the top will be absorbed by the tie rods of the anti-explosion floor, to limit the lifting of the rotating plugs to 800 mm. The quantity of sodium that will be ejected has been calculated as 27 kg. the reactor containment building has been designed for an overpressure of 250 mbar, resulting from the combustion of 100 kg of sodium, and will get isolated by signals from pressure and activity monitors. There is provision for controlled release of activity through the stack.

2.3 PROTOTYPE FAST BREEDER REACTOR

The PFBR is a 1,250 MWt/500MWe, sodium-cooled, pool-type, mixed oxide (MOX)-fueled reactor having two secondary loops. The primary objective of PFBR is to demonstrate techno-economic viability of fast breeder reactors on an industrial scale. The reactor power is chosen to enable adoption of a standard turbine as used in fossil-fuel power stations, to have a standardized design of reactor components resulting in further reduction of capital cost and construction time in the future and compatibility with regional grids. The design of FBTR core needed 85% enriched uranium oxide fuel, for which enriched uranium was not available at affordable prices on the international market. Hence, mixed carbide fuel of 70% PuC–30% UC was used. The PFBR being a commercial demonstration plant, a proven fuel cycle is essential. The mixed oxide (MOX) fuel (20% PuO_2 + 80% UO_2) is selected on account of its proven capability of safe operation to high burnup, ease of fabrication, and proven reprocessing.

Detailed studies were carried out to compare the advantages and disadvantages of pool and loop type designs. Better safety features of pool type—main vessel with no nozzles leading to high integrity of the vessel; relatively large thermal inertia leading to ease in design of decay heat removal and availability of more time for the operator to act; and large diameter of the main vessel with internals leading to significantly lower strain in the main vessel in case of core disruptive accident—led to selection of the pool-type design for a primary circuit configuration. The pool type concept also enables further extension of the design to larger power reactors in the future. The excellent performance of sodium components in fast breeder reactors obviates the need for easy maintenance on these components, which is possible in a loop-type reactor. A two-loop design has been adopted in view of its economic benefits and it meets the safety requirements (Figure 2.8). The reactor inlet and outlet temperatures are 397/547°C while steam conditions are 490°C, 166 kg/cm². The steam cycle selected is steam reheat with integrated steam generator (combined evaporator and superheater) instead of sodium reheat (separate evaporator and superheater sections) to simplify the design of the steam generator and associated circuits and for ease of operation. A dedicated safety grade decay heat removal system is provided in the reactor vessel, and it eliminated the necessity of designing secondary sodium and steam water systems as safety grade systems.

The reactor is designed with core sodium outlet temperature of 547°C, which is made possible by the use of SS 316 LN as structural material and capability to perform inelastic analysis for creep-fatigue damage assessment. Based on the choice of the steam generator material as ASTM Gr 91 (modified 9Cr–1Mo) and optimization studies, the steam parameters at turbine stop valve of 16.6MPa and 490°C have been fixed. The overall cycle efficiency is 40%.

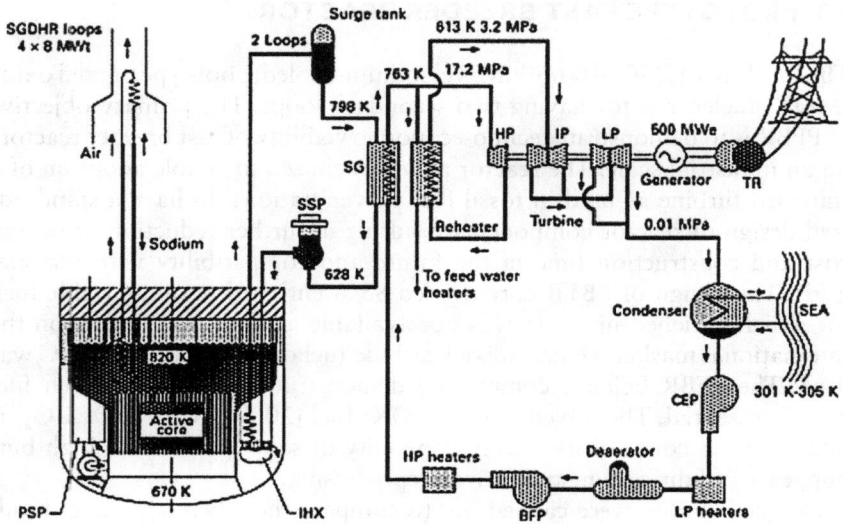

Figure 2.8 Schematic of PFBR with major components.

Source: Chetal et al., The Design of the Prototype Fast Breeder Reactor. *Nuclear Engineering and Design* 236, 2006.

2.3.1 Reactor Core

Figure 2.9 shows the core configuration. A core concept with two fissile enrichment zones of 21/28% PuO_2 is adopted for power flattening. The active core where most of the nuclear heat is generated consists of 181 fuel subassemblies. Each fuel subassembly contains 217 helium-bonded pins of 6.6-mm diameter. Each pin has a 1,000-mm column of annular MOX fuel pellets and 300 mm each of upper and lower blanket columns. The pin diameter of 6.6 mm is selected considering the Pu inventory, fuel fabrication cost, and breeding. The clad material used is 20%CW 15Ni–14Cr–2Mo + Si + Ti (D9). The maximum linear power in the fuel pin is 450 W/cm.

The initial peak fuel burnup is limited to 100 GWd/t due to deformation and swelling behavior of the D9 material used. In the long run, use of improved clad material of D9 (with addition of tramp elements phosphorus and boron) and use of ferritic steel of 9Cr–1Mo type for wrapper material are planned to achieve a targeted burnup up to 200 GWd/t. Spent subassemblies are stored for one campaign in internal storage with one-third of the active core being replaced during every fuel-handling campaign. Two rows of blanket subassemblies are provided surrounding the inner and outer fuel regions. Twelve absorber rods, i.e., nine control and safety rods (CSR) and three diverse safety rods (DSR), are arranged in two rings. Two independent and diverse shutdown systems are provided for ensuring a safe shutdown of the reactor even when one system is not available. Both systems are

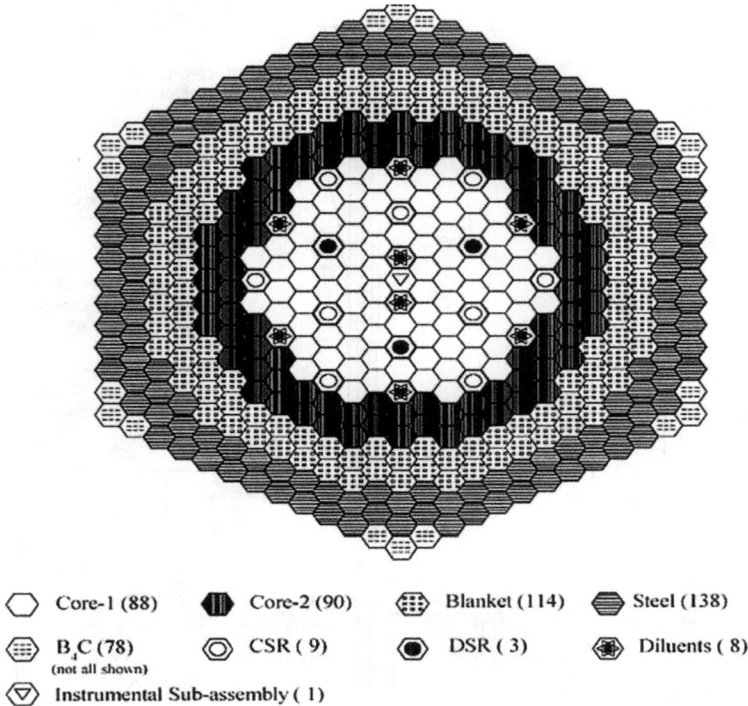

Figure 2.9 PFBR core schematic.

Source: Chetal et al., The Design of the Prototype Fast Breeder Reactor. *Nuclear Engineering and Design* 236, 2006.

designed to shut down the reactor in less than 1 s. In addition, axial shielding is provided within the subassemblies, and radial shielding subassemblies are provided within the core. These are optimized in order to have the required flux at in-vessel neutron detector locations. They also serve the purpose of limiting the activation of the secondary sodium, radiation damage of grid plate, and helium production in core cover plate.

2.3.2 Reactor Assembly

The entire primary sodium circuit is contained in a large-diameter vessel (Ø 12,900 mm), called the *main vessel*, and consists of core, primary pumps, intermediate heat exchanger, and primary pipe connecting the pumps and the grid plate (Figure 2.10). The vessel has no penetrations and is welded at the top to the roof slab. The main vessel is cooled by cold sodium to maintain it at lower temperatures and enhance its structural integrity. The core subassemblies are supported on the grid plate, which in turn is supported on the core support structure. A core catcher provided below the core support

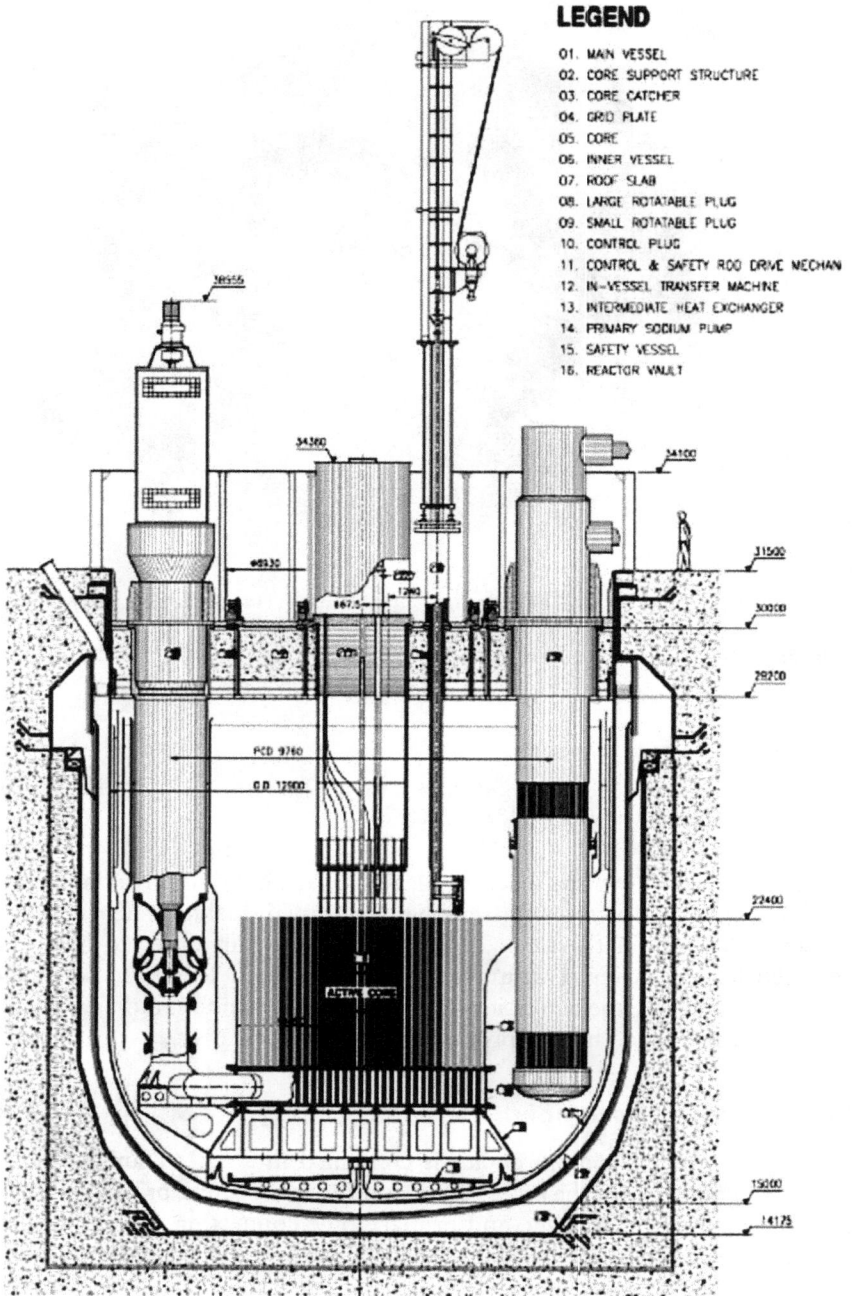

LEGEND

01. MAIN VESSEL
02. CORE SUPPORT STRUCTURE
03. CORE CATCHER
04. GRID PLATE
05. CORE
06. INNER VESSEL
07. ROOF SLAB
08. LARGE ROTATABLE PLUG
09. SMALL ROTATABLE PLUG
10. CONTROL PLUG
11. CONTROL & SAFETY ROD DRIVE MECHAN
12. IN-VESSEL TRANSFER MACHINE
13. INTERMEDIATE HEAT EXCHANGER
14. PRIMARY SODIUM PUMP
15. SAFETY VESSEL
16. REACTOR VAULT

Figure 2.10 PFBR reactor assembly.

Source: Chetal et al., The Design of the Prototype Fast Breeder Reactor. *Nuclear Engineering and Design* 236, 2006.

structure is designed to take care of a meltdown of seven subassemblies and prevents the core debris from encountering the main vessel. The main vessel is surrounded by the safety vessel, closely following the shape of the main vessel, with a nominal gap of 300 mm to permit robotic and ultrasonic inspection of the welds in the vessels. The safety vessel also helps keep the sodium level above the inlet windows of the intermediate heat exchanger, ensuring continued cooling of the core in case of a leak of the main vessel. The interspace between main and safety vessels is filled with inert nitrogen. An inner vessel separates the hot and cold pools of sodium. The main vessel is closed at its top by a top shield, which includes a roof slab, large and small rotatable plugs, and a control plug.

2.3.3 Main Heat Transport System

Liquid sodium is circulated through the core using two primary sodium pumps. The sodium enters the core at 397 °C and leaves at 547°C. The hot primary sodium transfers the heat to secondary sodium through four intermediate heat exchangers (Figure 2.8). The non-radioactive secondary sodium is circulated through two independent secondary loops, each having a sodium pump, two intermediate heat exchangers, and four steam generators. The choice of four steam generators per loop is based on overall optimization studies carried out considering capital cost, outage cost, and operation cost with three steam generators in the affected loop in case of a leak in one steam generator. The primary and secondary pumps are vertical, single-stage and single-suction centrifugal type, with variable-speed AC drives and are provided with flywheels to meet the flow-halving times of 8 s and 4 s, respectively. An AC pony motor of 30 kW rating is additionally provided for each of the primary pumps. The steam generator is a once-through, integrated-type design using straight tubes and an expansion bend in each tube. The decay heat is removed using the steam water system under normal conditions.

In case of off-site power failure or nonavailability of the steam water system, the decay heat is removed by a passive safety grade decay heat removal circuit consisting of four independent loops (Figure 2.11). Each safety grade decay heat removal loop is rated for 8MWt and consists of a decay heat exchanger immersed in the hot pool, one sodium/air heat exchanger with damper controls, and associated sodium piping and tanks. Diversity is provided for the decay heat exchanger, air heat exchanger, and dampers. The circulation of sodium and air is by natural convection. Crack opening of air heat exchanger dampers and sodium flow monitoring ensure poised condition for safety grade decay heat removal whose operation is automatic.

2.3.4 Steam Water System

The steam water system uses a reheat, regenerative cycle using live steam for reheating. High-pressure superheated steam from the steam generators drives

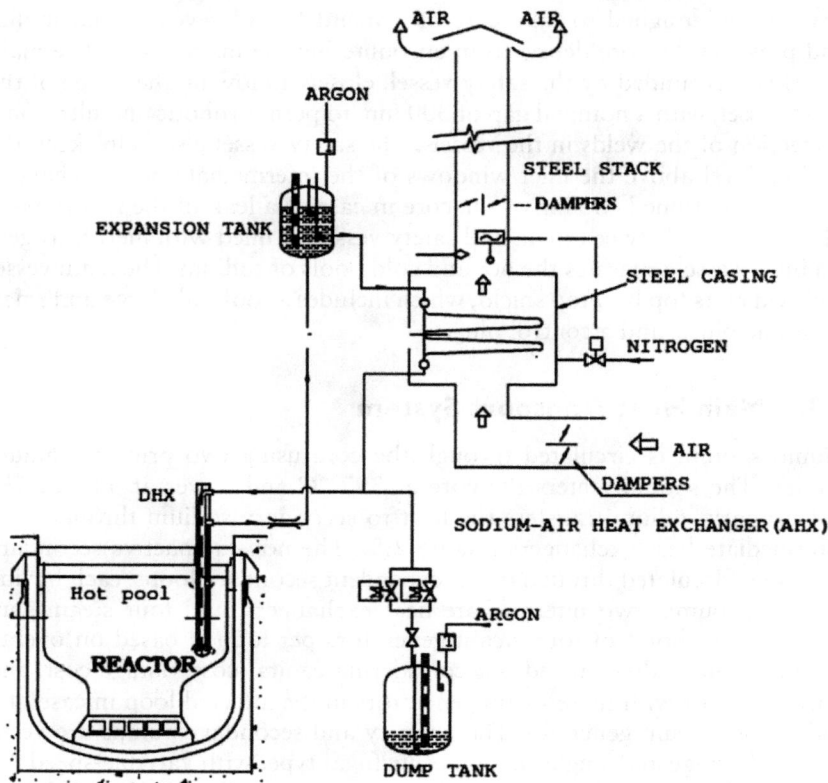

Figure 2.11 Safety grade decay heat removal loop.

Source: Raj et al., *Sodium Fast Reactors in Closed Fuel Cycle*, CRC Press, 2015.

a turbo alternator of 500-Mwe capacity. The turbine design is standard, used in fossil fuel–fired thermal power plants. Three 50% capacity boiler feed pumps are provided to deliver feed water to steam generators (Figure 2.8). Two of the pumps are turbo driven and the remaining one is motor driven. Feed water is heated in six stages, five in surface-type feed water heaters and one in a direct-contact deaerator. A steam separator is provided at the common outlet of the steam generator for start-up purpose. A turbine bypass of 60% capacity is also provided to take care of spurious turbine trips, when power is run down to 60%. The condenser is cooled by seawater in a once-through system. The condenser tubes are made of titanium.

2.3.5 Instrumentation and Control

The reactor power is controlled manually. The burnup compensation of reactivity is very small (22 pcm/d). Six fission chambers with a sensitivity of

1 cps/nv are provided above the core, at the bottom of a control plug, for safety considerations. Cover gas activity and delayed neutrons in the primary sodium are monitored for failed fuel detection systems. Sodium samples from each fuel subassembly outlet are taken using three selector valves for locating a failed subassembly. Two chromel–alumel thermocouples are provided to monitor the temperature of sodium at the outlet of each fuel subassembly. The flow delivered by the sodium pumps is measured using an eddy current flow meter, and safety action is taken on the power-to-flow ratio signal beyond the set limit. These provisions ensure that there are at least two diverse safety parameters to shut down the reactor safely for each design basis event that has impact on core safety.

SCRAM parameters from core monitoring systems and heat transport systems are connected to a plant protection system to automatically shut down the reactor in case any parameter crosses the limit. Steam generator tube leaks are detected by a leak detector (hydrogen in sodium) provided at the outlet of each steam generator module and an additional detector provided in the common outlet header. Two hydrogen-in-argon detectors are installed in the cover gas space of a surge tank. Acoustic leak detectors are also installed at various locations on the outer shell of the steam generator. Separate backup control room and fuel handling control rooms are also provided. Instrumentation directly concerned with reactor safety is designed using hardwired systems except core thermocouples, which are processed by real-time computers. Nonnuclear systems use a state-of-the-art distributed digital control system to take advantage of multiplexed signal transmission and reduced cabling, leading to cost savings. Safety signals are converted into digital form and are connected to the distributed digital control system for display in the control room.

2.3.6 Safety

To ensure safety, a design with adequate safety margin, early detection of abnormal events to prevent accidents, and mitigation of consequences of accidents, if any, is adopted. The reactor is designed with various engineered safety features such as two independent fast-acting diverse shutdown systems and decay heat removal systems with passive features of natural circulation of intermediate sodium/air, along with diversity in design of a decay heat exchanger and an air heat exchanger. A core catcher to collect molten fuel and containment is provided as defense in depth for beyond design basis events. In the event of a total instantaneous blockage of a single subassembly, fuel melting starts, and the melting progresses at the most to the neighboring six subassemblies before the reactor is shut down by delayed neutron detection system. The molten fuel and debris are collected in a suitably dispersed manner to avoid recriticality and ensuring long-term cooling. The amount of molten fuel released in the melting of seven subassemblies is only 0.3 t. On the other hand, calculations indicate that 1 t of fuel in the most reactive configuration is required to achieve recriticality. Therefore, the slumped molten fuel of seven subassemblies does not have any recriticality potential. Despite this, a core catcher with provision to collect the full molten core without recriticality is provided below the core support structure as a defense-in-depth measure. A chimney is provided at the center of the core catcher to aid natural convection flow of sodium.

Selection of design features, detailed design analysis, and rigorous manufacturing specifications minimize the risk of sodium leaks from components, piping, and leaks resulting in sodium–water reaction in steam generator. Nevertheless, provisions have been made for early detection of sodium leaks and sodium–water reaction in the steam generator and safety actions to minimize the consequence of the leaks. Additionally, the design also provides for in-service inspection of the main and safety vessels, secondary sodium piping, and steam generator.

All the sodium piping inside the reactor containment building is provided with a double envelope with nitrogen inerting to avoid sodium fire. The structural integrity of primary containment, intermediate heat exchanger, and decay heat exchanger is assured under core disruptive accident, which results in an energy release of 100 MJ, the theoretically assessed upper bound value for energy release. A rectangular, single, non-vented, reinforced concrete containment designed for 25 kPa is provided. The maximum pressure inside the containment is estimated to be 13 kPa with the conservative assumption of instantaneous burning of all the sodium that is ejected above the roof slab under a core disruptive accident. The containment is designed such that the dose limits at the site boundary for design basis accident of 100 mSv is not exceeded under a core disruptive accident.

2.4 NEUTRONIC CHARACTERISTICS OF SFRs

From a neutronics perspective, among the unique features of SFRs that have safety implications are their compact core size, operation with a fast neutron spectrum, and utilization of Pu and higher-enriched uranium (relative to thermal reactors) in the fuel. Table 2.1 shows the comparative neutronic features of a PWR and SFR.

The following characteristics are relevant:

- Fast fission cross-sections are a few hundred times lower than for thermal fissions, requiring a higher concentration of fissionable fuel in a fast spectrum core.
- Smaller loss by parasitic capture in fuel and lesser poisonous effects from fission products lead to the possibility of higher fuel burnup and lower excess reactivity requirements for SFRs.
- With high burnups, good fission gas retention or venting is a consideration in SFR fuel system design.

Table 2.1 Comparison of typical PWR and SFR characteristics

Parameter	Thermal reactor	Fast reactor
Fissile Enrichment	~3–5% U235	10–30% Pu239/U235
Average Neutron Energy (ev)	~0.025	~100,000
Peak Burnup MWd/ton	~40,000	~150,000
Neutron Flux (n/cm²s)	10^{14}	$5-10 * 10^{15}$
Average Core Power density (W/cm²)	~100	~300–400
Average Fuel Specific Power (kW/kg)	~40	~100

- Fuel burnup in SFRs is usually limited not by reactivity but by radiation damage to the fuel pins (e.g., swelling).
- The possibility of leakage of dense hydrogenous (moderating) material into an SFR core must be avoided because of concern over prompt criticality brought on by positive reactivity associated with the softening of the neutron spectrum. To this end, all SFRs have the intermediate circuit, unlike PWRs, preventing leakage of water into the primary sodium circuit.
- SFRs generally have short prompt neutron lifetimes, and this needs to be considered in the design of the reactivity control system.
- In SFRs the effective delayed neutron fraction (βeff) is impacted negatively by Pu_{239} (β for Pu_{239} is only 0.00215 compared to 0.0068 for U_{235}) and positively by fast fission of the fertile U_{238} ($\beta = 0.0158$).
- The presence of a harder neutron spectrum in metal-fueled SFRs leads to significantly smaller Doppler feedback than in ceramic-fueled reactors.
- Bowing of fuel assemblies due to high radial temperature gradients across the core can lead to reactivity changes.

In view of a shorter neutron lifetime, there is a misconception regarding the requirement of very fast neutron absorber systems for the control of a chain reaction maintained by fast neutrons in SFR. The solution to the reactor kinetic differential equation with the consideration of delayed neutrons represented in Figure 2.12 provides better clarity. The period shown on the ordinate is the time during which the power of the reactor increases by a factor e, if appropriate reactivity (>1) is inserted without considering any feedback effects. The reactor would then attain the uncontrolled state. Under such uncontrolled situation, the SFR has a shorter period, compared to a PWR. However, under the controlled state of the reactor (reactivity introduction <1), the periods are practically the same for both SFR and PWR and are not distinguishable from each other from the reactor period point of view. The periods are so long that control is quite possible with technically simple regulating units.

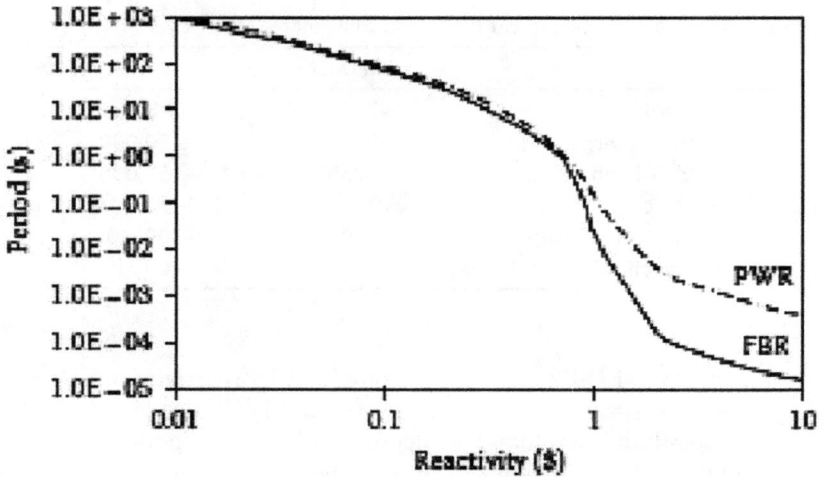

Figure 2.12 Period versus reactivity in PWR and fast reactors.

Another misconception about the lower delayed neutron fraction is that the safety margins available for the control are highly restricted in an SFR. As opposed to this view, fast reactors are more stable even with a small number of delayed neutron fractions. Variations in the operating parameters—for example, inlet temperature, coolant flow, and power—have considerably less influence on the reactivity in the case of a SFR. This may be obvious in the case of reactivity changes, which are produced due to a change in the reactor inlet temperature of around 1 K, which is ~0.0015\$/K compared to 0.01–0.1\$/K for a PWR depending on the burnup. In addition, a short shutdown time (<1 s) is easily possible in a SFR due to the short core height (one of the favorable consequences of high-power density).

2.5 THERMAL-HYDRAULIC CHARACTERISTICS OF SFR

From a thermal-hydraulics perspective, among the unique features of SFRs that have safety implications are their compact core size of relatively high-power density and the use of low-pressure sodium as primary coolant. The following characteristics are relevant:

- Liquid metals such as sodium (and potassium) have relatively low melting temperature (98°C) and high boiling temperature (~900°C). They remain in liquid form over a wide range of temperatures.
- A pool-type reactor coupled with a low-pressure primary system makes the occurrence of a large loss-of-coolant accident (LOCA) unlikely.

- Operating at low pressures, sodium will not completely flash on depressurization.
- For liquid metals, the Prandtl number, Pr, is less than 1 and the convective heat transfer coefficient (given in the Nusselt number, Nu) is a function of the Peclet number, Pe = Re * Pr where Re is the Reynolds number. Generally, for Pe < 100, heat transfer is dominated by conduction (Nu does not vary much with Pe) and is not affected by the coolant velocities.
- The relatively large mass of sodium in a pool-type reactor (versus a loop-type) not only provides large heat capacity to dampen temperature rise in off-normal transients but also influences the control and load following characteristics of the overall heat transport systems.
- No moderator in a fast spectrum core leads to a more compact core, that is, higher power density and higher specific power. This translates to more restrictive coolant flow passages and more severe heat removal requirements.

High specific powers and power densities require large heat transfer areas and high heat transfer coefficients to be used to reduce fuel centerline and cladding temperatures and avoid melting. To this end, experience with sodium so far has been very good.

ASSIGNMENT

1. Present the important features of the fast breeder test reactor (FBTR).
2. "FBTR can operate at full power even if turbine is not available." Is this statement true? Give reasons.
3. What are the important features of a prototype fast breeder reactor (PFBR)?
4. Compare loop-type and pool-type SFRs.

REFERENCES

Bhoje S.B., et al., (1985), Design and construction of the Fast Breeder Test Reactor. In: Proceedings of the IAEA International Symposium on Fast Breeder Reactors, Lyons, France.

Chetal S.C., et al., (2006), "The design of the Prototype Fast Breeder Reactor", *Nucl. Eng. Des*, Vol. 236, pp. 852–86: doi:10.1016/j.nucengdes.2005.09.025.

Hemanath M.G., et al., (2007), "Cellular convection in vertical annuli of fast breeder reactors", *Ann. Nucl. Energy*, Vol. 34, pp. 679–686: doi:10.1016/j.anucene.2007.03.004.

Patil R.K. and Shimjith S.R., (2014), Applications of Digital Reactivity Meter based on Kalman Filtering Technique in Indian Nuclear Reactors, BARC Newsletter, Issue NO. 336, 2014.

Raj B., Chellapandi, P. and Rao, P. V. (2015), *Sodium fast reactors with closed fuel cycle*, CRC Press.

Srinivasan G., Suresh Kumar K.V., Rajendran B. and Ramalingam P.V. (2006), "The fast breeder test reactor—design and operating experiences", *Nucl. Eng. Des*, Vol. 236, pp. 796–811: doi:10.1016/j.nucengdes.2005.09.024.

Suresh Kumar et al., (2011), "Twenty-five years of operating experience with the fast breeder test reactor", *Energy Proc*, Vol. 7 pp. 323–332: doi:10.1016/j.egypro.2011.06.042.

Chapter 3

Reactor Heat Transfer

3.1 INTRODUCTION

The heart of any nuclear reactor is the core where heat is generated due to the fission process. Liquid sodium (Na) picks up the heat from the fuel assemblies and transports to the upper plenum, which is a sodium capacity where sodium streams from different fuel assemblies at different temperatures mix. The sodium at the mixed mean temperature enters the intermediate heat exchanger and comes out into the cold pool or lower plenum of the reactor assembly. From the lower plenum sodium is pumped back into the core via an inlet plenum or grid plate. This chapter deals with modeling of reactor kinetics and thermal modeling of the core, upper and lower plenums, and inlet plenum of the reactor.

3.2 REACTOR CORE

The heart of the SFR is the core where heat is generated due to fission. The core contains fuel, reflector, blanket, and shielding subassemblies. All are hexagonal in shape for effective utilization of space (Figure 3.1). The fuel subassemblies comprise closely packed fuel rods. The fissile material in the reactors have been either U_{235} or a mixture of U_{235} and Pu_{239}. To control the fission reaction, absorber rods comprising Boron 10 enriched to about 60% to 90% are used. Normally these absorber rods are distributed uniformly over the core to have effective control.

The neutron energy level needs to be as high as possible to achieve maximum breeding, pointing to the need to use structural materials with minimum absorption for neutrons. There is also the need to have as high a steam temperature as possible to have higher power cycle efficiency, which would result in low fuel inventory and consequently lower waste generation. For this the core outlet temperature must be as high as possible. This again points to the need for structural materials with enough fatigue and creep strength at high temperatures. The most important objective is low cost of power generation for which the overall fuel cycle costs must be minimized.

DOI: 10.1201/9781003283188-3

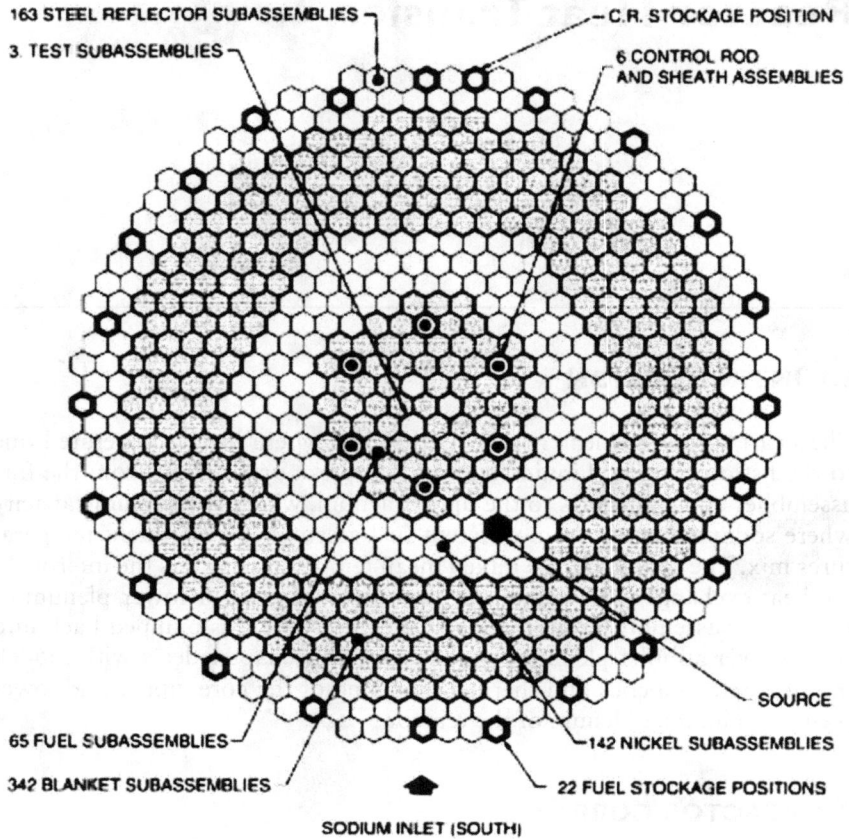

163 STEEL REFLECTOR SUBASSEMBLIES
3. TEST SUBASSEMBLIES
C.R. STOCKAGE POSITION
6 CONTROL ROD AND SHEATH ASSEMBLIES
SOURCE
65 FUEL SUBASSEMBLIES
142 NICKEL SUBASSEMBLIES
342 BLANKET SUBASSEMBLIES
22 FUEL STOCKAGE POSITIONS
SODIUM INLET (SOUTH)

Figure 3.1 FBTR core layout.

Source: Srinivasan et.al, The Fast Breeder Test Reactor—Design and Operating Experiences. Nuclear Engineering and Design 236, 2006.

The fuel used must have the capability to withstand exposure to radiation for a long time. With this characteristic it is possible to extract higher energy per unit mass of fuel (MWd/t). This is referred to as the burnup capability of fuel. The longer life also helps in minimizing the number of fuel-handling campaigns, improving plant availability.

3.2.1 Core Description

The fuel subassemblies are surrounded by blanket subassemblies. Blanket subassemblies comprise fertile materials like U_{238} or $Thorium_{232}$. These absorb a neutron and get converted to Pu_{239} and U_{233}, which are fissile materials. The blanket subassemblies in turn are surrounded by steel subassemblies that act as neutron reflectors and help keep the loss of neutrons from the core to a minimum. Then we have the shielding subassemblies that are made up of borated steel. These absorb the neutrons leaking past the reflector subassemblies and reduce the neutron flux seen by the reactor vessel that houses all the subassemblies. All subassemblies have many cylindrical pins (Figure 3.2) arranged in a triangular pitch. The individual pins comprise stainless steel tubes referred to as cladding inside which the fuels, fertile, control materials in pellet form are packed. The pins are put together in a hexagonal sheath called hexcan (Figure 3.2). The number of pins in a subassembly varies based on the power, reactivity worth, etc.

The main candidate fuel materials are metal, oxide, carbide, and nitride of uranium-plutonium. Of these metal and oxide fuels have been used in the prototype fast reactors. Thorium-uranium fuel can also be used, but higher breeding ratios are achievable only with uranium-plutonium. Early US and UK experimental fast reactors used the metal fuel. The neutron energy spectrum in metal fuels is high due to the absence of moderation from oxygen atoms that are present in oxide fuels, but they have a lower melting point (~1,150°C) compared to oxide fuels (~2,750°C). The higher the energy, the better is the breeding, and metal fuel has the highest breeding ratio.

Mixed oxide fuel of U-Pu has been used in nearly all large plants. This is essentially due to the high burnup capability. Also there exists a large amount of data on this fuel in view of experience with a large number of water reactors, not only in fabrication techniques but also in reprocessing. The softer neutron spectrum due to the presence of two oxygen atoms attached to U-Pu reduces the breeding. With mixed carbide and nitride, the neutron spectrum is harder than oxide due to presence of only one C/N atom. However, sufficient database and experience are needed before taking up these fuels on a commercial scale. Hence, the option to be pursued with preference to breeding is the metal fuel, where some experience exists.

3.2.2 Fuel Pin

The overall length of the fuel pin is determined by the active lengths of core, upper and lower axial blankets, fission gas plenum, etc. The governing factors in arriving at the length are pressure drop, coolant temperature rise, and sodium void worth.

The fissile enrichment required to attain criticality in a fast reactor is considerably higher than that required for thermal reactors, due to the low fission cross-sections for all fuels at higher neutron energies. Pu_{239} content typical of SFR is 12% to 30% depending on the reactor size and configuration. The fissile fraction in a fast reactor is four to five times that in thermal reactors. The fuel pin diameter is a compromise between the breeding ratio and the fissile inventory. Pin diameters are in the range of 6.6 to 8 mm in most FBRs.

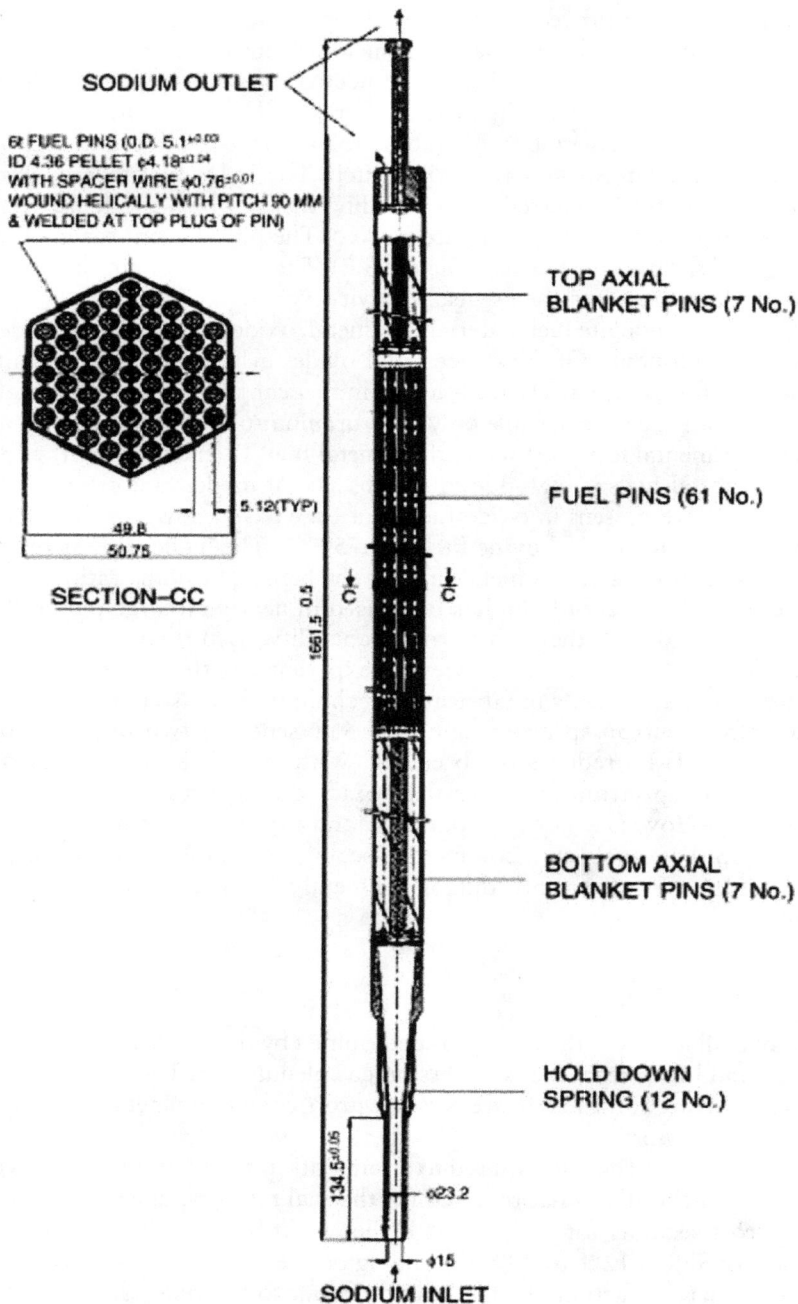

SODIUM OUTLET

61 FUEL PINS (O.D. 5.1⁺⁰·⁰³
ID 4.36 PELLET φ4.18±⁰·⁰⁴
WITH SPACER WIRE φ0.76±⁰·⁰¹
WOUND HELICALLY WITH PITCH 90 MM
& WELDED AT TOP PLUG OF PIN)

49.8
50.75

5.12(TYP)

SECTION–CC

TOP AXIAL
BLANKET PINS (7 No.)

FUEL PINS (61 No.)

1661.5 ⁰·⁵

C   C

BOTTOM AXIAL
BLANKET PINS (7 No.)

HOLD DOWN
SPRING (12 No.)

134.5⁺⁰·⁰⁵

φ23.2

φ15

SODIUM INLET

Figure 3.2 FBTR fuel subassembly.

Source: Srinivasan et.al, The Fast Breeder Test Reactor—Design and Operating Experiences. *Nuclear Engineering and Design* 236, 2006.

3.2.3 Subassembly

The subassembly (hexcan) serves the following purpose:

- provide structural support for the pins
- directs the flow toward the pins
- allows individual orificing of subassemblies
- provides a barrier to potential propagation of fuel failure.

The number of pins in a subassembly is governed by many considerations including its reactivity worth, criticality during shipment, decay heat removal, etc. Each fuel subassembly houses thin fuel pins in a triangular pitch. The FBTR fuel subassembly is 4 to 5 m long and weighs ~250 kg. Each subassembly consists of an orifice at the inlet to admit sodium flow proportional to its heat generation.

3.3 COOLANT SELECTION

The requirements for coolant are:

- Minimum neutron moderation
- High heat transfer coefficient
- Minimum pumping power
- Low activation
- Compatibility with structural materials
- Low cost

Water is excluded from moderation consideration, but gases and liquid metals (alkali metals) are candidates as coolant. Mercury, though used in the Clementine reactor, has not been considered further because of the toxicity. NaK is a liquid at room temperature and will spread throughout an area in the event of a leak, whereas sodium tends to solidify on cold surfaces. NaK will spontaneously ignite in air much more easily than sodium. For these reasons, NaK is considered somewhat more hazardous to handle in the event of a leakage. Hence, the idea of NaK as coolant was abandoned. Sodium has been used in most reactors, but since its M.P. is ~98°C, the components need to be preheated before filling liquid sodium.

The attention to the new coolants has arisen due to the exothermic reaction of sodium with air (in case of a pipe leak) and water (in case of a leak in an SG tube). New coolants that have been considered for the fast breeder reactors are lead and lead-bismuth eutectic (LBE). Interest in these coolants stems from the fact that LBE was used in Russian nuclear submarines. Both lead and LBE have got higher boiling points (1,743°C and 1,670°C, respectively) than sodium and are thus quite suitable. Besides, lead and LBE have

Table 3.1 Heat Transfer Data for Various Coolants (Vaidyanathan, 2013)

Property	Na	NaK'	Hg	Pb	H₂O	Pb-Bi
T_{melt} (°C)	98	18	−38	328	0	125
T_{boil} (°C)	880	826	357	1,743	100	1,670
c_p (kJ/kg. °C)	1.3	1.2	0.14	0.14	4.2	0.143
K (W/m. °C)	75	26	12	14	0.7	14
h (W/m². °C)†	36,000	20,000	32,000	23,000	17,000	22,500
Relative Pumping Power	0.93	0.93	13.1	11.5	1.0	6.29

got superior neutronic properties and do not have any reaction with water. Nevertheless, physical interaction with water can induce deleterious effects on structures. Also, both these fluids have higher densities (almost 10 times that of sodium), which will mean a much greater pumping power requirement. There is lack of proper database regarding their compatibility with structural materials. Table 3.1 compares the properties of the different coolants.

3.4 CONTROL MATERIAL SELECTION

Boron in the form of boron carbide (B_4C) is used as absorber rod material for controlling the power of the nuclear reactors. Natural boron contains 20% B_{10}, and the rest is the B_{11} isotope. B_{10} is the isotope with neutron absorption properties. While water reactors use natural boron, fast reactors use boron enriched in B_{10} to ~60–70%. Boron swells on absorption of a neutron due to the production of helium. This necessitates replacement of control rods once in ~2 years. Tantalum is being considered as a possible substitute, primarily because of its favorable swelling characteristics and availability. The disadvantage is the 115-day half-life gamma decay from Ta_{182} to W_{182}, which causes long-term decay heat removal problems. Also, Ta is soluble in sodium. Europium oxide is another candidate. It has twice the neutron absorption capability compared to boron. However, it has many disadvantages from reactivity and low thermal conductivity considerations. In light of this, boron is continued to be used as an absorber material in control rods.

3.5 STRUCTURAL MATERIAL SELECTION

The fuel clad is the first structure whose integrity is important. It provides physical separation of fuel from coolant and prevents fission gas from entering the primary sodium. The hexcan supports all the pins of a subassembly. The requirements of material for both clad and hexcan are similar. They are:

- High temperature strength
- Irradiation resistance
- Low neutron absorption
- Ease of manufacturing
- Compatibility with coolant and fuel

Zircalloy used in water reactors has not been favored due to low strength at higher temperatures. Austenitic stainless steel, normally cold-worked to 20%, has been used in many reactors. Ferritic steels are also being considered for future-generation reactors.

3.6 HEAT GENERATION

The time-dependent fission power is calculated by solving the space-averaged one-energy group reactor kinetics equation. The one group energy assumption is reasonable, particularly for a fast reactor. The space-averaged model is quite adequate, since the core of an SFR responds, due to relative smallness of the core and the large neutron migration area, to changes in reactivity more uniformly than a light water reactor core.

Laureau et al. (2017) carried out a comparison of the space time kinetics (SK) and point kinetics (PK) models for a 600-MWe fast reactor cooled by sodium for the case of an unprotected loss-of-flow accident (ULOF). The PK and SK modeling showed a very good agreement. The difference on the reactivity maximum variation was limited to 3 pcm (percent milli). Thus, PK model can be used for dynamic simulation of transients in SFR. Fast reactor safety codes to date have therefore generally employed point kinetics; however, the reactivity feedback terms due to reactor temperature variations and material motion generally contain spatial effects.

The reactor power generation rate in the core depends on the external and feedback reactivities as depicted in Figure 3.3.

The PK equations written in terms of the prompt neutron generation time may be expressed as:

$$\frac{dN}{dt} = \frac{\rho(t) - \beta}{\wedge} N(t) + \sum_i^m \lambda_i C_i \qquad (3.1)$$

$$\frac{dC_i}{dt} = \frac{\beta_i}{\wedge} N(t) - \lambda_i C_i \qquad (3.2)$$

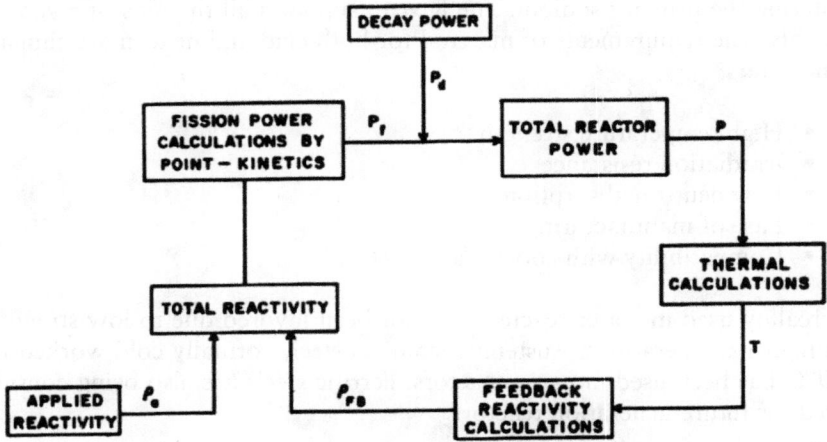

Figure 3.3 Scheme of reactor power calculation.

where:
N = neutron density
ρ = total reactivity
βi = effective delayed neutron fraction of ith group
β = total delayed neutron fraction
$\lambda\, i$ = decay constant of the ith group
C_i = precursor density of ith group of delayed neutrons
$\wedge$ = prompt neutron generation time
t = time (s)

By rewriting N and Ci,

$$n(t) = \left(\frac{N(t)}{N(0)} \right) \tag{3.3}$$

$$C_i(t) = \frac{C_i(t)}{C_i(0)} \tag{3.4}$$

where $C_i(0) = \dfrac{\beta_i}{\wedge \lambda_i} N(0)$

Equations (3.1) and (3.2) become

$$\frac{dn}{dt} = \frac{\rho - \beta}{\wedge} n(t) + \frac{1}{\wedge} \sum \lambda_i C_i \tag{3.5}$$

Table 3.2 Delayed Neutron Half-Lives and Decay Constants

Group Index	U_{235} λ_i (S^{-1})	β_i	Pu_{239} λ_i (S^{-1})	β_i	Typical SFR Fuel λ_i (S^{-1})	β_i
1	0.0127	0.00063	0.0129	0.00024	0.0129	0.000082
2	0.0317	0.00351	0.0311	0.00176	0.0312	0.000776
3	0.115	0.00310	0.134	0.00136	0.133	0.000666
4	1.311	0.00672	0.331	0.00207	0.345	0.001354
5	1.40	0.00211	1.26	0.00065	1.41	0.000591
6	3.87	0.00043	3.21	0.00022	3.75	0.000181
	Total	0.0165	Total	0.0063	Total	0.00365

and

$$\frac{dc_i}{dt} = \lambda_i \left[n(t) - c_i(t) \right] \tag{3.6}$$

The values of λ_i and β_i for U_{235} and Pu_{239} are given (Table 3.2) for six groups of delayed neutrons. The total reactivity ρ at any time is the sum of the applied external reactivity ρ_{cr} (e.g., control rod movement) plus the sum of various reactivity feedback contributions ρ_{fb},

$$\rho[t] = \rho_{cr} + \rho_{fb} \tag{3.7}$$

It should be noted that the reactivity effects are inherently space dependent. This is not only because the temperatures vary spatially; even for the same temperature, the magnitude of the effect will depend on the location within the reactor.

3.7 REACTIVITY FEEDBACK

The following contributions are reported in Hummel and Okrent (1970) with respect to feedback reactivity (Figure 3.4).

- Doppler
- Sodium density and voiding
- Fuel axial expansion
- Structural expansion
- Bowing
- Fuel slumping (in case of a meltdown)

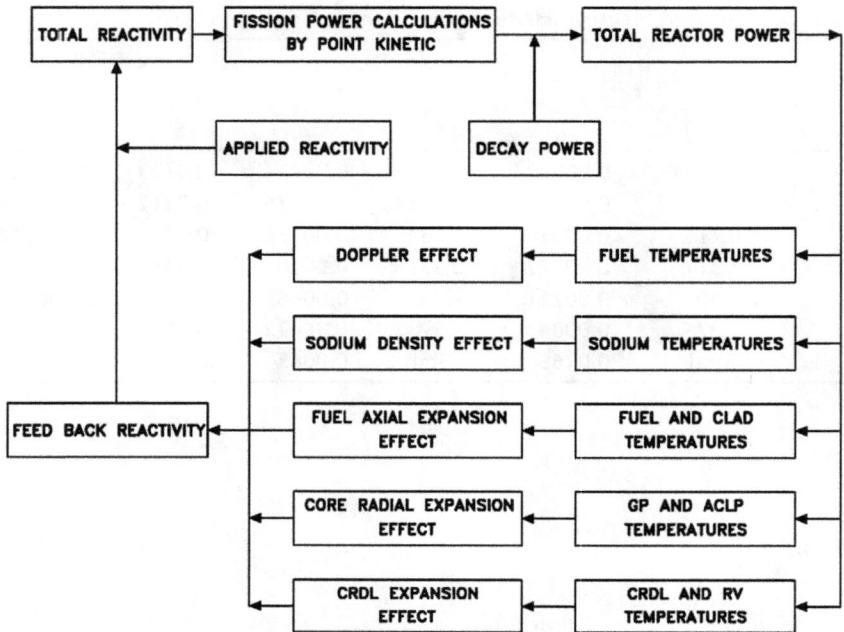

Figure 3.4 Reactivity feedback calculation scheme.

3.7.1 Doppler Effect

It is important to have a prompt negative reactivity feedback that reverses a power transient if the reactor becomes prompt critical. Mechanical action by control rods is too slow after prompt criticality is reached. A prompt negative feedback is particularly important for fast reactors because two mechanisms, fuel compaction and sodium loss, have the potential to make the reactor super prompt critical (KAERI, 1998). The Doppler effect is the most important and reliable prompt reactivity effect in current thermal and fast reactor designs, which utilize highly fertile U_{238} material concentrations. For fertile materials U_{238} or Th_{232}, the predominant reaction is the absorption of neutrons, and hence the Doppler effect in these materials increases the absorption of neutrons and reduces the core reactivity, i.e., the feedback is negative.

The Doppler effect is due to the increased kinetic motion of fuel atoms, as measured by an increase in fuel temperature, resulting in the broadening of cross-section resonances and increased resonance absorption. The Doppler reactivity comes predominantly from the capture of low-energy neutrons. Ceramic-fueled reactors, due to the presence of oxygen or carbon in the fuel, have a soft enough neutron spectrum to have a large Doppler effect. Both oxide- and carbide-fueled fast reactors possess much higher Doppler coefficients than metal-fueled reactors. This represents one of the important advantages of ceramic over metal fuel.

The Doppler coefficient is defined as the change in multiplication factor k, associated with an arbitrary change in the absolute fuel temperature. Since this coefficient is found to vary as inverse of fuel temperature, a temperature-independent doppler parameter can be defined as

$$\left(\frac{d\rho}{dT}\right)_{DOP} \equiv \frac{1}{k}\frac{dk}{dT} \cong \alpha_{DOP}T^{-1} \tag{3.8}$$

$$(\Delta\rho)_{DOP} = \alpha_{DOP} \ln\left(T_2/T_1\right) \tag{3.9}$$

where
T_1 = initial reference fuel temperature K
T_2 = current fuel temperature K

3.7.2 Sodium Density and Void Effects

Heating of the sodium coolant decreases sodium density and can ultimately lead to vaporization (voiding). The sodium void/density reactivity effect is exceedingly space dependent. Sodium void/density from the center of the core yields a highly positive reactivity effect, and sodium loss from near the edge gives a negative effect. The density decrease has two competing effects: increased leakage, which results in negative reactivity; and spectral hardening (increase in average neutron energy level) due to decrease of sodium, which is a positive effect. The net effect is reactor specific.

3.7.3 Fuel Axial Expansion Effect

The role of axial expansion of fuel in the normal solid fuel pin geometry is to provide a prompt negative reactivity feedback at the start of a power transient. This mechanism is the principal prompt negative feedback available in a metal-fueled fast reactor. The lack of a strong Doppler coefficient for metal fuels, which is caused by the characteristic hard neutron spectrum, has historically been compensated for by the presence of a reliable axial expansion coefficient.

Fuel axial expansion results in an increase in active core height and a decrease in fuel density, leading to a net decrease in reactivity. Actual mechanisms for expansion are difficult to model, especially for ceramic fuels.

Fuel pellet cracking or friction between the pellet and inner clad will affect expansion. In case the fuel pellets are not stacked in perfect contact within the clad, there may be negative effect due to fuel pin settling.

3.7.4 Structural Expansion

As the core heats up, there is the radial expansion of the fuel clad, hex-can, and the core support structures (grid plate), which tend to effectively increase the pitch-to-diameter ratio of the fuel lattice, reducing reactivity. This effect is a slow one, as structural capacity is large. This feedback is very important for long-term unprotected transients (*unprotected* means failure of reactor shutdown system). Also, there is expansion of the control rod due to rise in hot plenum sodium temperatures. In case of top supported reactor vessels, this effect would be negative. For the case of FBTR during an unprotected loss of flow, Figure 3.5 gives the feedback reactivity contributions from the different elements, namely control rod expansion, sodium expansion, and grid plate expansion during this event (Vaidyanathan et al., 2010). The grid plate expansion is initially positive due to initial fall in reactor inlet temperature. Later, this feedback becomes negative due to a rise in temperature and contributes significant negative feedback.

3.7.5 Bowing

This is due to the differential thermal expansion of subassemblies because of temperature gradients (Figure 3.6). Core-flowering is a type of bowing, and it means that one subassembly expands and induces stresses on the

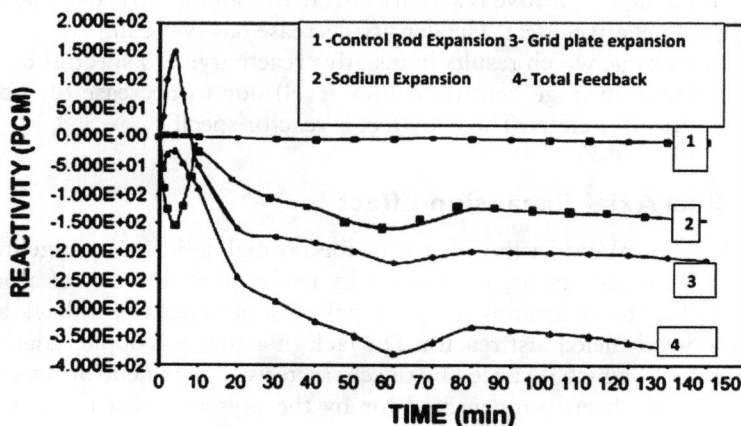

Figure 3.5 Reactivity contributions for ULOF in FBTR.

Source: Vaidyanathan et al., Dynamic Model of Fast Breeder Test Reactor. *Annals of Nuclear Energy* 37, 2010.

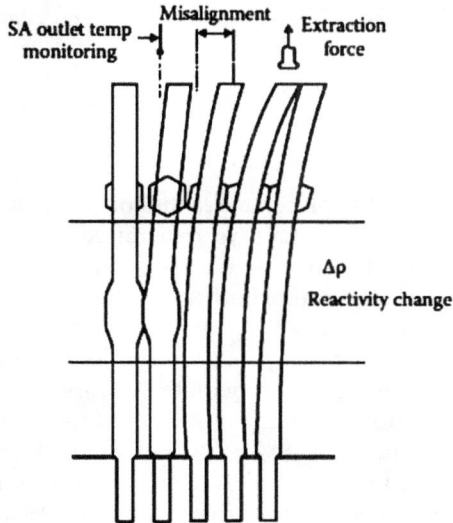

Figure 3.6 Bowing profile of a cluster of subassemblies.

Source: Raj et al., *Sodium Fast Reactors in Closed Fuel Cycle*. CRC Press, 2015.

surrounding subassemblies, causing the core to expand in a radial direction. The result of core extension is displacement of the subassemblies in the core, leading to an increase of the gap between the units. When a fuel subassembly bows toward the center, the effect is positive, and vice versa. This has a large response time associated with it. It has a reactivity effect, and this was observed in the French reactor PHENIX (Filip Gottfridsson, 2011).

At the end of 1980s, PHENIX encountered, while operating at full power, an earlier unexperienced phenomenon that led to an automatic shutdown of the reactor. The signal of the neutron chambers registered very rapid oscillations with high amplitudes. During the period between 1989 and 1990, PHENIX suffered from this event four times. These transients were named *Arrêt d'urgence par réactivité négative* (AURN). In English this means automatic emergency shutdown by negative reactivity. The events occurred while operating at or close to full power; the first three at 580 MWth and the last one at 500 MWth. AURN were all detected by the neutron chambers, which are located beneath the reactor vessel to measure the neutron flux. This phenomenon only lasted for several hundreds of milliseconds before the reactor was shut down automatically by the control rods. The control rods were triggered by the first reactivity drop, since the amplitude of the drop went below the threshold for negative reactivity transient. After almost two years of investigation, a complete explanation of the phenomenon was not found, though the most probable cause was the radial movement of the subassemblies. Furthermore, in the safety analysis, which was based on all plausible scenarios that could cause AURN, it was concluded that the safety of the reactor was not affected.

3.8 DECAY HEAT

When a nuclear fission takes place, several fission products are formed. Also, neutrons, alpha, beta, and gamma rays are emitted. Most of the heat generated is instantaneously absorbed by the fuel and, to a lesser extent, in the coolant and structural material. The fission fragments are unstable (have excess neutrons) and undergo several transitions by beta emissions before reaching stability. Each of these processes results in heat generation, and this heat must be removed. Depending on the history of the core—i.e., duration of power production, power level, and isotopic content of fuel—the decay heat can be as much as 6–8% of the power at which the reactor operated before shutdown. Figure 3.7 shows the fission product decay heat vs. time curve for a typical fast reactor after shutdown. Decay heat data for all nuclides are available through data files such as evaluated nuclear data file (ENDF/BIV) available from Brookhaven National Laboratory (Rose and Burrows, 1976).

When a reactor is tripped or shut down, the fission reactions stop, but fission fragments continue to generate heat as explained above. Hence, a nuclear reactor must have long-term cooling after shutdown to remove decay heat.

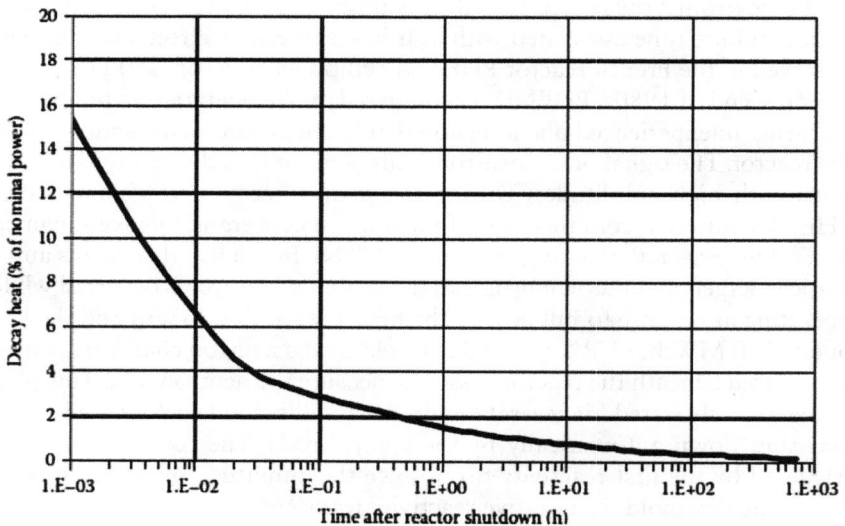

Figure 3.7 Fission product decay heat curve.

Source: Raj et al., *Sodium Fast Reactors in Closed Fuel Cycle*. CRC Press, 2015.

An approximation for the decay heat curve valid from 10 s to 100 days after shutdown is given as

$$P / P_o = 0.066\left((\tau - \tau_s)^{-0.2} - \tau^{-0.2}\right)$$

where:

P is the decay power

P_0 is the reactor power before shutdown

τ is the time since the reactor starts, s

τ_s is the time of reactor shutdown measured from the time of start-up, s.

3.9 SOLUTION METHODS

Point kinetics equations are stiff differential equations due to large variation between the time constants of n and C. Computational solution through conventional explicit method will give a stable consistent result only for smaller time steps, due to the small value of Λ (~6 * 10^{-7}s). For ensuring numerical stability, the time step sizes required would be approximately 100 to 1,000 times Λ, i.e., 0.0006 to 0.00006 s. However, with smaller time steps, the rounding of error will affect accuracy. The truncation error is a matter of concern for larger time steps. Smaller time steps would result in large computation time for a plant simulation code. Hence, various methods have been developed.

There has been a great deal of research done in the field of reactor kinetics. Much of the work has focused on eliminating the problem of stiffness. Chao and Attard (1985) investigated this problem and developed the "stiffness confinement method" that places emphasis on the physical analogies of the PK equations in order to deal with the problem. Sanchez (1989) devised and implemented an A-Stable Runge-Kutta Method to avoid the problem of stiffness. Several other creative schemes have been implemented as well. da Norbrega (1971) uses a Pade approximation to the solution, and Aboanber and Nahla (2002) derived a technique based on the analytical inversion of polynomials to aid in the solution to these equations. Important ones generally applicable to transient analysis of fast reactors are briefly presented below.

3.9.1 Prompt Jump Approximation

The simplest method is termed *prompt jump approximation*, wherein due to low value of Λ, the term $\Lambda(dn/dt)$ is neglected. It means that any disturbance in reactivity affects power instantaneously. It is reported that solutions using this method match closely with exact solution if the total reactivity ρ is less than 0.5 β. For most transients the reactivity additions are small, as the reactor would be tripped much before the reactivity approaches β. The set point for reactivity trip for most reactors is given by $d\rho/\rho = 10^{-4}$, i.e., 0.02β.

3.9.2 Runge Kutta Method

Fourth-order Runge kutta method is commonly used to solve simultaneous differential equations (Leonoapiders and John Seinfield, 1971). They possess the advantage that they are self-starting and do not require previous values or higher derivatives of the dependent variable. However, the estimation of the error is not simple. A very small value of prompt neutron lifetime Λ, combined with the necessity to perform subtractions yielding small differences between large numbers, limits the time step to $\sim 10^{-4}$ s. Thus, for classes of problems where variables change slowly with time, the number of computations becomes enormous.

3.9.3 Kaganove Method

This is a polynomial method, where it is assumed that over an integration time step, the parameters n and ρ may be represented by a second-order polynomial (Kaganove 1960).

$$n(t) = n_0 + n_1.t + n_2.t^2 \tag{3.10}$$

$$\rho(t) = \rho_0 + \rho_1.t + \rho_2.t^2 \tag{3.11}$$

where n_0 and ρ_0, respectively, are the initial values of n and ρ and n_1, n_2, ρ_1, and ρ_2 are constants to be evaluated. Substituting for $n(t)$ and $\rho(t)$ in Equations (3.5) and (3.6) yields equations relating n_1, n_2 to ρ_1, ρ_2. Imposing the boundary conditions for ρ at $\Delta t/2$ and t, we can solve for n_1 and n_2. During the time interval Δt, $\rho(t)$ is considered independent of $n(t)$.

3.9.4 Comparison of the Different Methods

A comparison of the methods for thermal reactors has been reported by Charagi and Thangasamy (1976). Here the neutron lifetime is of the order of 10^{-3} s as compared to 10^{-7} s in fast reactors. It was seen that the Runge kutta method is most promising for accidental situations where the reactivity changes over a few seconds. In a fast-spectrum nuclear reactor, the neutron generation time ranges from 50 μs to 1.0 ms, while in a thermal-spectrum reactor, it is much larger in magnitude, ~ 10 ms. Owing to these very short time scales, conventional explicit integration methods, such as Euler's and higher-order Runge-Kutta, require a highly restrictive (short) time steps to avoid exponential amplification of the error, with unreasonably large computation time for convergence (Greenspan et al., 1968). The prompt jump approximation was found to be the most economical computationally without the loss of accuracy for all types of transients. Hence the prompt jump approximation has been used in most of the studies.

3.9.5 Solution Methodology

The solution method adopted is the prompt jump approximation. In the following lines details of the technique are presented. As a first step, the kinetics equations are rewritten in terms of power P_n, which is proportional to neutron density N as

$$\frac{dP_n}{dt} = \left\{ \frac{\rho - \beta}{\wedge} \right\} P_n + \sum_{i=1}^{6} \lambda_i C_i \tag{3.12}$$

$$\frac{dC_i}{dt} = \frac{\beta_i}{\wedge} P_n - \lambda_i C_i \ for \ i = 1, 2, \ldots, 6 \tag{3.13}$$

where $\beta = \sum \beta_i$ and $C_i = \beta_i Pno/(\wedge\lambda i)$

The delayed neutron fractions β_i and fission product precursor decay constants λ_i are supplied as data.

The transient solution is obtained by prompt jump approximation for power P_n, i.e., $(\wedge dP_n/dt = 0)$, and by simple finite difference in time the equations for precursors concentration Ci as

$$\left\{ \frac{\rho - \beta}{\wedge} \right\} P_{n_j} + \sum_{i=1}^{6} C_{i_j} = 0 \tag{3.14}$$

and

$$\left\{ \frac{C_{i_j} - C_{i_{j-1}}}{\Delta t} \right\} = \frac{\beta_i}{\wedge} \left\{ P_{n_j} - P_{n_{j-1}} \right\} 0.5 - 0.5 \lambda_i \left(C_{i_j} - C_{i_{j-1}} \right) \tag{3.15}$$

where subscript j refers to the time index.

The above is a set of seven simultaneous equations that are then solved to obtain the following expression for P_{nj} and C_{ij}:

$$P_{nj} = \frac{\sum_{i=1}^{6} \lambda i \left[\frac{\beta_i \Delta t P_{nj-1}}{(2 + \lambda i \Delta t) \wedge} + C_{ij-1} \left(\frac{2 - \lambda i \Delta t}{2 + \lambda_i \Delta t} \right) \right]}{\left(\frac{\rho - \beta}{\wedge} \right) + \sum_{i=1}^{6} \left[\frac{\lambda_i \beta i \Delta t}{\wedge (2 + \lambda_i \Delta t)} \right]} \tag{3.16}$$

$$C_{ij} = \frac{\beta_i \Delta t \left(P_{nj} + P_{nj-1} \right)}{\wedge \left(2 + \lambda_g \Delta t \right)} + C_{ij-1} \left(\frac{\left(2 - \lambda_i \Delta t \right)}{\left(2 + \lambda_i \Delta t \right)} \right) \, for \, i = 1,2,\ldots\ldots,6 \qquad (3.17)$$

From this, the total power P_{th} is obtained after adding the power due to decay of fission products as follows:

$$P_{thj} = P_{nj} + P_{dj}$$

where decay power P_{dj} is generally fitted in the form

$$P_d / P_{tho} = At^B + C$$

where P_{tho} is the steady-state total power of the reactor and A, B, and C are empirical constants that are obtained to fit the fission product heat generation data available from reactor physics calculation.

The total feedback reactivity is calculated as follows:

$$\rho_{RG} + \rho_{CR} + \rho_{Na} + \rho_{fu} + \rho_c + \rho_{Dop} + \rho_{Ext}$$

where ρ_{RG} is the feedback reactivity due to radial expansion of the grid plate, ρ_{CR} due to control rod expansions, ρ_{Na} is due to volumetric sodium expansion inside the fuel and blanket portion of the core, ρ_c is due to the axial clad steed expansion, ρ_{fu} is due to axial fuel expansion, and ρ_{Dop} is due to Doppler effect brought about by the changes in fuel temperature. All these are functions of respective temperature differences from the steady state, ΔT. The calculations of these individual effects are as follows:

$$\tau_{RG} \frac{dT_{RG}}{dt} + T_{RG} = T_{RI}, \Delta T_{RG} = T_{RG} - T_{RGO}$$

$$\rho_{RG} = X_{RG} \Delta T_{RG}$$

where T_{RG} is the current grid plate temperature, T_{RG0} is the grid plate temperature at t = 0, τ_{RG} is the structural time constant based on the thermal capacity coolant heat transfer coefficient, T_{RI} is the reactor inlet temperature, and X_{RG} is the reactivity coefficient data obtained from reactor physics calculation. The reactivity effect due to control rod expansion is modeled in a similar manner.

For the core portion involving 10 zones, the reactivity feedback is represented by

$$\rho_m = \sum_{r=1}^{10} \left[X_{mri} \Delta T_{mri} \right] for \, m = Na, fu, c$$

where T_{mri} are the core temperatures obtained from subassembly heat transfer calculations and X_{mri} is the reactivity coefficient. The summation indices r and i represent, respectively, the radial and axial positions of the finite difference grid.

Doppler effect for 10 radial and 14 axial zones is calculated from

$$\rho_{Dop} = \sum_{r=1}^{10}\sum_{i=1}^{14}\left[\left(k_d \ln\frac{T}{T_0}\right)_{ri}\right]$$

where T_{ri} and T_{ori} are fuel temperatures in degrees Kelvin and k_{dri} are the Doppler constants. The value of ρ_{Ext} is specified as data input to simulate any control rod movement.

3.10 HEAT TRANSFER IN PRIMARY SYSTEM

The heat transfer model for the primary sodium system comprises the core thermal model and the hot and cold plenum models.

3.10.1 Core Thermal Model

Figure 3.8 shows a schematic of the fuel and the associated coolant channels of the fast breeder test reactor (FBTR), India. Almost all the heat generated in the core is produced in the fuel and blanket assemblies. The heat generated in the control and other shielding and reflector assemblies is insignificant. The simplest representation of the core is by a point core model. In this model, all the heat generation is approximated by a single lumped point. This model may be sufficient when the balance of the plant is being analyzed in detail. The next approach would be to represent the core by a suitably averaged channel. In this method the axial temperatures of fuel, clad, and coolant are computed for an average channel. Temperature distributions in other subassemblies are obtained by suitable weighting factors. This model is an improvement over the point model, as the temperature distribution would help in a more accurate treatment of reactivity feedback effects. Further improvement can be obtained by representing each assembly through an assembly averaged channel.

The average fuel pin temperature T in an assembly can be represented by the heat conduction in cylindrical coordinates as given below

$$\rho c\frac{\partial T}{\partial t} = \frac{1}{r}\frac{\partial}{\partial r}\left(kr\frac{\partial T}{\partial r}\right) + \frac{\partial}{\partial r}\left(k\frac{\partial T}{\partial z}\right) + q''' \tag{3.18}$$

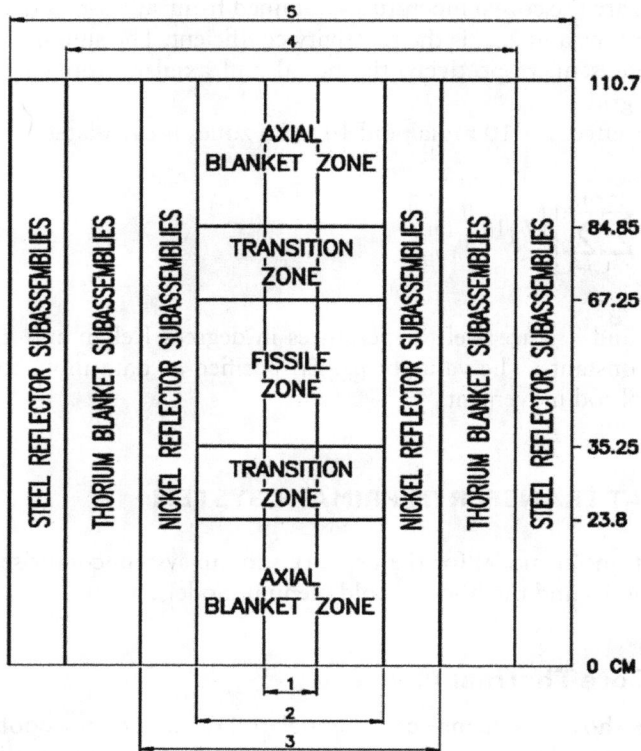

Figure 3.8 Core model for dynamic simulation of FBTR.

where q''' is the volumetric heat generation rate, c is the specific heat, ρ is the density, and k is the thermal conductivity of fuel. The above model of average pin neglects the intra-channel cross-flow effects.

A further simplification is possible if the axial conduction term can be neglected. In the case of oxide fuel, the temperature drop between the centerline and fuel surface is about 1,500°C compared with 150°C across the full core. Thus, radial conduction is dominant.

3.10.2 Fuel Restructuring

Mixed oxide fuel is known to undergo densification or restructuring when irradiated (Neimark et al., 1972). The void in the center of the fuel element is clearly visible, as well as the very long grains associated with the columnar region and the larger grains of the equiaxed region. Outside the equiaxed grain region and adjacent to the cladding is an annulus of fuel with the original microstructure. The temperatures in this region are too low to cause any observable restructuring of the fuel material.

A major impact of this densification is improvement in fuel thermal conductivity (Neimark et al., 1972). It is reported that fuel restructuring takes place within a short time of operation. With improvement in conductivity, the fuel center line temperatures would come down. Hence, thermal conductivity data as a function of temperature is important. To consider the effect of varying thermal conductivity with temperatures, the fuel is divided into concentric rings, and the conduction equations are solved for each ring with appropriate thermal conductivity values.

3.10.3 Gap Conductance

The gap conductance between the fuel pellet and the cladding also plays an important role in the estimation of temperatures in the fuel. The magnitude of the gap conductance depends on the bonding between the fuel and the clad. With oxide fuel, the gap is filled with helium and fission gases released during fission. With an increase in burnup the gap may get closed, i.e., the fuel touches the clad. Studies have been carried out to arrive at a simple correlation for gap conductance that could be used in numerical simulation studies of oxide fuels (Lavarenne et al., 2019). The following is the correlation,

$$h_{gap}(LHR,B) = 10^{-2}(3+LHR)\left[1+tanh(2\times10^{-5}\times B)\right]$$

where:
h_{gap} is fuel-clad gap conductance $W.cm^{-2}.K^{-1}$
LHR is linear heat rating, $kW.m^{-1}$
B is burnup, $MWd.\, t^{-1}$

This correlation captures the main physical phenomena where gap conductance increases linearly with LHR as an increase in local power leads to thermal expansion of the fuel, which reduces the gap size, and to an increase of the gas temperature and thermal conductivity.

3.10.4 Fuel Thermal Model

Calculations were carried out with fuel divided into many concentric rings, and it was seen that a two concentric ring model is quite adequate for plant dynamics (Govindarajan and Vaidyanathan, 1985). The mathematical model of a fuel pin treated as two concentric rings and clad together with coolant can be written as:

$$(MC)f_1\frac{\partial Tf_1}{\partial t} = .5P_{Th} - h_1(Tf_1 - Tf_2) \tag{3.19}$$

$$(MC)f_2 \frac{\partial Tf_2}{\partial t} = .5P_{Th} + h_1\left(T_{f1} - T_{f2}\right) - h_2\left(T_{f2} - T_{Na}\right) \tag{3.20}$$

$$(MC)_{Na} \frac{\partial T_{Na}}{\partial t} = h_2\left(T_{f2} - T_{Na}\right) - QC_{Na}\frac{\partial T_{Na}}{\partial z} \tag{3.21}$$

where T_f is the temperature of fuel ring, M is mass of the fuel ring, C is the specific heat of fuel, h_1 is heat conductance between inner ring and outer ring, h_2 is heat conductance between outer ring and sodium, Q is the sodium flow, and T_{Na} is the temperature of sodium. Heat conductance depends on conductivity. Subscripts f_1, f_2, na refer to inner fuel ring, outer fuel ring, and sodium, respectively. For additional rings of fuel, equations can be formed in a similar manner. The sodium heat transfer coefficient is high enough that the temperature drop across the boundary layer between sodium and clad is only ~10–15°C. Thus, sodium and clad are treated as a single region.

3.10.5 Solution Technique

Each fuel pin is divided into a few axial zones, and the above equations are applied to each zone. The steady-state solution is obtained by setting the time derivatives to zero, replacing the spatial derivative by backward difference, and solving the set of algebraic equations obtained by Gaussian elimination.

For the dynamic regime, the following finite difference approximations are made:

$$T = \left(T_{i,j} + T_{i,j-1}\right) * 0.5$$

$$\partial T / \partial t = \left(T_{i,j} - T_{i,j-1}\right) / \Delta t$$

where i, j represents space and time.

The set of algebraic equations obtained after the above substitutions are solved for each zone, by Gaussian elimination.

3.11 DETERMINATION OF PEAK TEMPERATURES: HOT SPOT ANALYSIS

The actual temperatures of sodium/clad and fuel could be higher because of the uncertainty in the calculations for power and uncertainty in the properties of fuel, clad, and sodium, besides measurement uncertainties. For each of the design variables there exists some pessimistic values, and if temperatures are calculated using these values, the designer can be quite sure that nowhere else will the temperature be superior to this critical temperature.

Three methods for analysis are deterministic method, statistical method, and semi-statistical method. The deterministic method is the oldest and most conservative method. The statistical method is an optimistic method, since all variables that appear in calculation are not randomly distributed. Here, it is assumed that the variables follow the law of standard statistical distribution. In a semi-statistical method, the variables that cause the hot spot temperature are separated into two principal groups, that is, variables of statistical origin and nonstatistical origin. The nonstatistical ones are treated cumulatively while the statistical parameters are treated statistically.

For purposes of system simulation code, a single hot spot factor is calculated for the given design and used as input data in the calculations. The hot spot clad temperature is calculated as

$$T_{hs} = T_{Nai} + F\left(T_{Nao} - T_{Nai}\right)$$

where T_{hs} is the hot spot temperature, T_{Nai} and T_{Nao} are, respectively, the inlet and maximum outlet temperature across the hottest channel, and F is the hot spot factor. The factor is calculated for steady state and considered constant during the transient state.

3.12 CORE THERMAL MODEL VALIDATION IN FBTR AND SUPER PHENIX

The FBTR reactor power was increased from 7.2 MW to 7.7 MW in 7 s by withdrawal of a control rod. At the end of 7 s the reactor got tripped by a drop of all control rods. The drop of all control rods simultaneously into the core is referred to as safety control rod accelerated movement (SCRAM). Figure 3.9 shows the central subassembly temperatures predicted by DYNAM along with plant measurement (Vaidyanathan et al., 2010). It is seen that there is a good match, validating the thermal model for forced flow conditions.

In October 1966, a blockage of one of the subassembly inlets occurred at the Enrico Fermi fast reactor in Detroit, USA, resulting in coolant boiling, fuel failure, and activity transport outside primary circuit (APDA, 1966). To guard against such an event, all FBR designs have multiple holes for coolant entry. Even if 50% of the holes get blocked, the flow reduction would be only about 10%. In addition, all the fissile subassemblies have thermocouples at the outlet to detect the rise in temperatures consequent to a plugging. The method consists in measuring the outlet sodium temperature from the subassembly and comparing it with the mean value of a group of certain similar subassemblies.

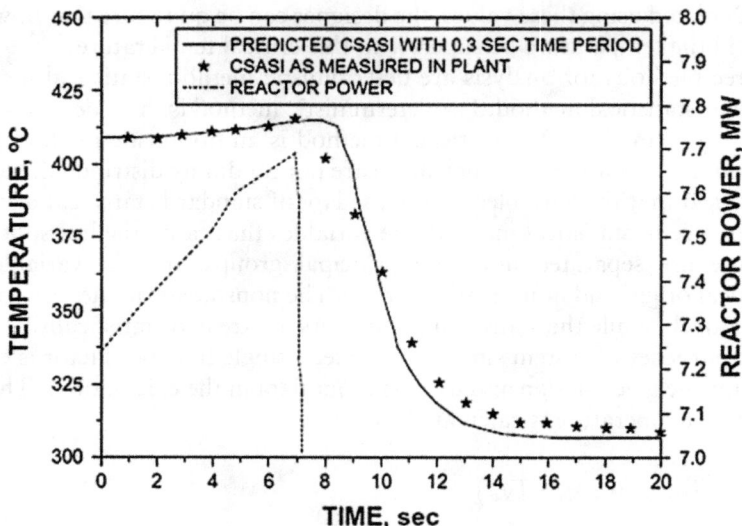

Figure 3.9 Central subassembly outlet temperature.

Source: Vaidyanathan et.al, Dynamic Model of Fast Breeder Test Reactor. *Annals of Nuclear Energy* 37, 2010.

Plugging detection capability of the SUPER PHENIX reactor fuel subassembly has been reported in literature (Gourdon, 1979; Maurin, 1982). It is indicated that the plugging resulting in a step reduction in flow of 27% can be detected before hot spot clad crosses 800°C, and a 70% reduction can be detected before maximum sodium temperature crosses 900°C. The DYNAM code core thermal model was run for this case based on the data available in literature (Vaidyanathan et al., 2010). It was seen that a step reduction in flow of 32% could be detected before hot spot clad crosses 800°C, and a 71% step reduction in flow could be detected before the maximum sodium temperature crossed 900°C. Considering the uncertainty in data, the comparison can be considered satisfactory. This gives a good validation of the thermal model.

3.13 MIXING OF COOLANT STREAMS IN UPPER PLENUM

The upper plenum or hot pool receives the exit flows of sodium at different temperatures from all the subassemblies. Besides the mixing process of all streams that takes place in the hot pool, the hot sodium also exchanges heat with the immersed structures, such as the control plug, and with cover gas. The heat transfer to cover gas is small to be considered for a systems analysis. In pool-type reactors, the hot pool exchanges heat through the inner vessel wall as well as with the cold pool.

Mixing phenomena in the upper and lower plenums of the reactor needs to be accurately modeled, as it has a large response time associated with it. In many cases the mixing coefficients are obtained from experiments as a function of sodium flow through the core or through multidimensional codes. Studies carried out with two- and three-dimensional modeling of the hot pool under steady-state conditions have revealed that there is good mixing in the hot pool and one-dimensional mixing model is applicable (Natesan, 2020). Under transient conditions like reactor SCRAM and primary flow coast-down, the cooler sodium from the fuel and other assemblies enters the hot pool at lower velocities. This brings into the picture the role of hot sodium at the top of the pool with layers of colder sodium in the layers below. This is referred to as *thermal stratification*. Thermal stratification leads to axial thermal stresses in the structures in contact with sodium and needs to be considered in the mechanical design. Studies carried out for such a case indicate differences between one-, two-, and three-dimensional models as shown in Figure 3.10. It shows the sodium inlet temperature to IHX, which is the output of the hot pool mixing.

A three-dimensional model represents the thermal stratification in a hot pool more accurately compared to other models due to the more realistic representation of the hot pool geometry. But the computational effort involved for this model is enormously high compared to other models. The one-dimensional predictions deviate from the three-dimensional model

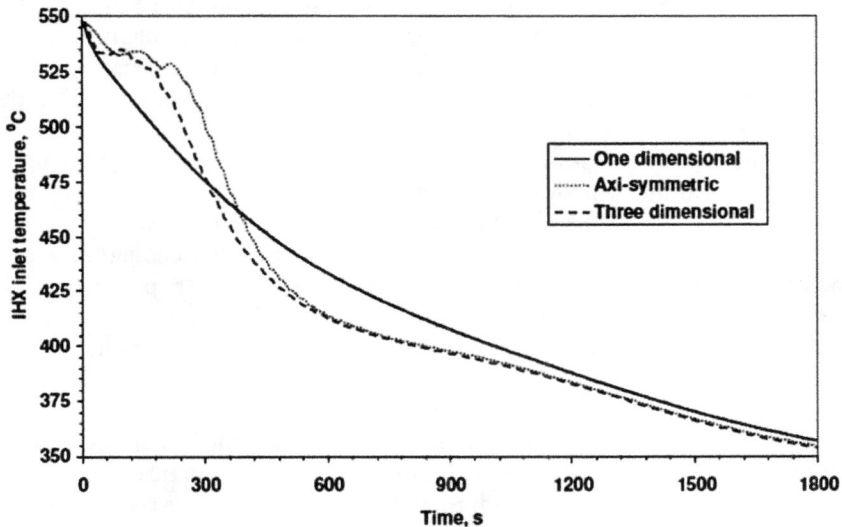

Figure 3.10 IHX sodium inlet temperature evolution in PFBR during SCRAM.

Source: Natesan K., Development of mathematical models and investigation of FBR plant behaviour during coupled single and multi-dimensional approach, Doctoral Thesis, Homi Bhabha National Institute, Mumbai, India, 2020.

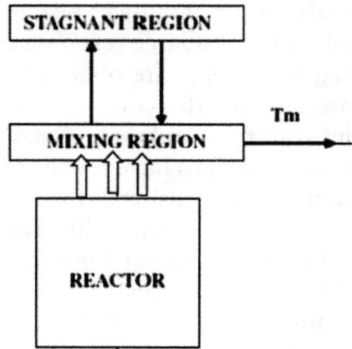

Figure 3.11 Upper-plenum model FBTR.

predictions by ~17 %. The deviation gets reduced to ~7 % with two-dimensional or axisymmetric representation of the hot pool. Considering the closeness of predictions between the two-dimensional and three-dimensional models, the axisymmetric modeling can be considered as a good approximation to represent thermal stratification effects in the hot pool. Plant dynamics simulations can be carried out with due consideration of thermal stratification effects in the hot pool by incorporating a 2D-based mixing model for the hot pool. However, the trend of temperature variations appears to be well predicted by the one-dimensional model. To improve the capability of the one-dimensional mixing model, the upper plenum can be treated in different zones with heat conduction between the capacities. A similar approach was used in the modeling of the dynamics of CRBRP in the USA (Yang and Agrawal, 1976).

Such an approach was adopted for FBTR in India. Sodium capacity up to the top of the outlet pipe was treated as a perfect mixing capacity in thermal contact with a stagnant sodium above it as shown in Figure 3.11.

Results obtained with such a model and comparison with actual measurements on the reactor for a loss of offsite power with reactor trip (Figure 3.12) show the applicability of such a model for system-wide transients. However, if one is interested to have more accurate predictions from the point of thermal loading of hot pool structures, a two-dimensional model of the hot pool could be integrated with a one-dimensional system code.

In the following the one-dimensional model used in the Indian fast reactor program is presented. The thermal capacities of the structures dipping in sodium are clubbed with the sodium thermal capacity in the two zones. Equations below model this process:

$$\left(MC\right)_m \frac{dT_m}{dt} = \sum_{i=1}^{n} Q_i C_{na}\left(T_i - T_m\right) - h\left(T_m - T_{st}\right) \qquad (3.22)$$

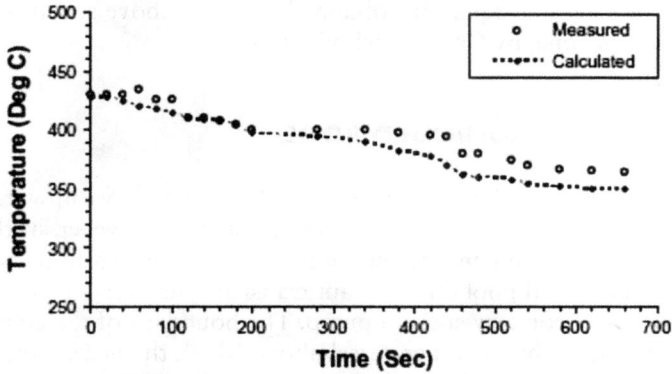

Figure 3.12 FBTR reactor outlet temperature—loss of off-site power.

Source: Vaidyanathan et.al, Dynamic Model of Fast Breeder Test Reactor. *Annals of Nuclear Energy* 37, 2010.

$$(MC)_{st} \frac{dT_{st}}{dt} = h(T_m - T_{st}) \tag{3.23}$$

Equation (3.22) is for the mixing volume where sodium streams from n channels mix, and Equation (3.23) is for stagnant volume. T_m is the temperature of the mixing region blow reactor outlet pipe, T_{st} the temperature of stagnant sodium above the mixing region, Q_i sodium flow from the ith core flow zone, MC is the mass capacity, and h the equivalent heat transfer coefficient between mixed and stagnant zones arrived at based on comparison between one-dimensional and multidimensional models. Subscripts m, na, and st refer to mixed mean zone, sodium, and stagnant zone.

3.13.1 Solution Technique

The steady-state solution is obtained by setting the time derivatives to zero, replacing the spatial derivative by backward difference, and solving the set of algebraic equations obtained by Gaussian elimination. For the dynamic regime, the following finite difference approximations are made:

$$T = (T_{i,j} + T_{i,j-1}) * 0.5$$

$$\partial T / \partial t = (T_{i,j} - T_{i,j-1}) / \Delta t$$

where i and j refer to spatial node and the time step, respectively.

The set of algebraic equations obtained after the above substitutions are solved for each zone, by Gaussian elimination.

3.14 LOWER PLENUM/COLD POOL

In loop-type reactors the sodium exiting the pump flows upward to the cold pool and there are not many other structures. However, in the pool-type reactors significant multidimensional temperature patterns would be developed in the cold pool during plant transients, namely reactor SCRAM and trip of one secondary sodium pump. The boundary of the cold pool at the top is occupied by the inner vessel above which the hot sodium pool is present (Figure 3.13). Sodium velocity at the outlet of IHX has downward direction. Therefore, sodium flow entering the upper region of the cold pool is very small.

During transients, the temperature of sodium at the outlet propagates initially to the lower region of the cold pool. Temperature of the upper region of the cold pool changes slowly due to diffusion rather than by

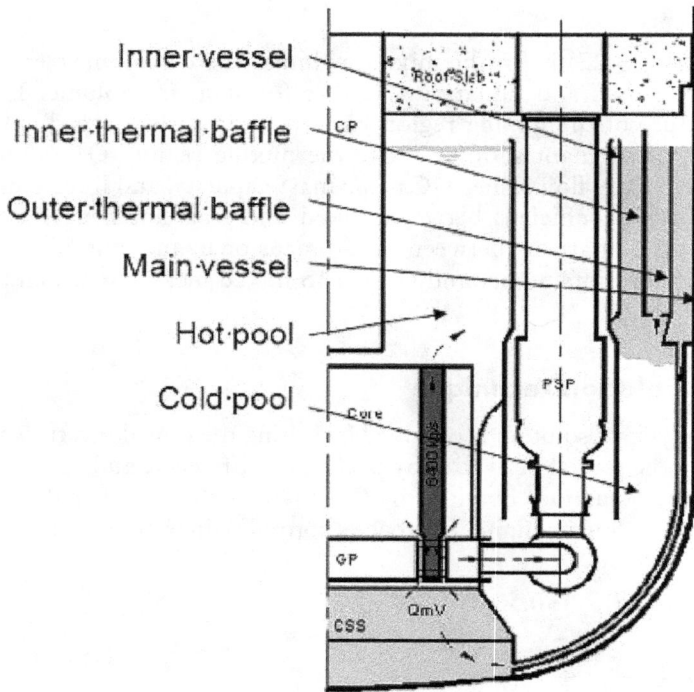

Figure 3.13 PFBR cold pool.

convection (due to low velocity of sodium prevailing in this region of the pool during transients). Other important transients that can cause imperfect mixing of coolant in the cold pool are trip of one secondary sodium pump or loss of cooling of steam generators connected to one loop. Transient thermal hydraulic analysis of the cold pool during reactor SCRAM and one secondary sodium pump trip events have been carried out using a three-dimensional model of the cold pool, and the predicted transient behavior has been compared against one-dimensional as well as multi-zone modeling of the cold pool (Natesan, 2020). In the one-dimensional model, the entire cold pool is represented by a perfect mixing model. In the multi-zone model applicable for pool-type SFR, the pool is divided into different sectors.

Temperature evolutions at the inlet of the pump predicted by the three-dimensional, single zone and multi-zone models for PFBR are shown in Figure 3.14. It can be observed that the multi-zone model predictions are close to that predicted by the three-dimensional model. Nevertheless, the one-dimensional model results are reasonably close from a system point of view.

For PFBR, in case of a secondary sodium pump trip in loop 1, there is sufficient temperature difference between the primary sodium at IHX outlet

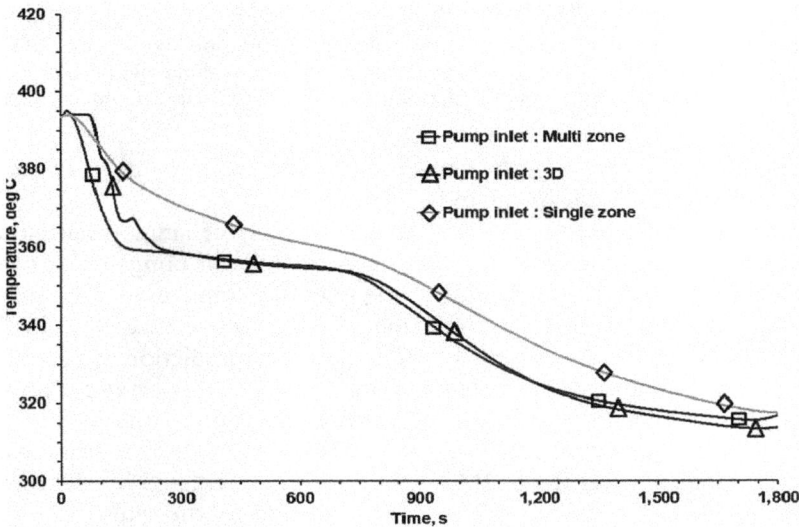

Figure 3.14 Temperature evolution at the inlet of pump predicted by various models during spurious SCRAM.

Source: Natesan K., Development of mathematical models and investigation of FBR plant behaviour during coupled single and multi-dimensional approach, Doctoral Thesis, Homi Bhabha National Institute, Mumbai, India, 2020.

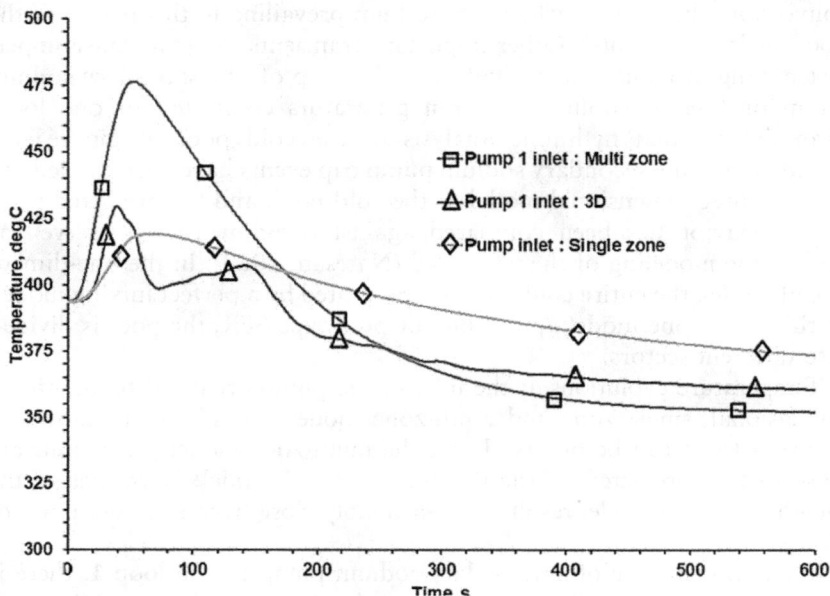

Figure 3.15 Predicted evolution of sodium temperature at inlet of pump I during one secondary pump trip.

Source: Natesan K., Development of mathematical models and investigation of FBR plant behaviour during coupled single and multi-dimensional approach, Doctoral Thesis, Homi Bhabha National Institute, Mumbai, India, 2020.

of the two loops. Results of the temperature evolution at the primary pump inlets are given in Figures 3.15 and 3.16.

It can be seen from Figures 3.15 and 3.16 that the single-zone model underpredicts the temperature evolution at the inlet of pump 1 and over-predicts the temperature evolution at the inlet of pump 2 compared to the predictions by the three-dimensional model. There is overprediction by the multi-zone model at the pump 1 location and underprediction at the pump 2 location. From the reactor safety point of view, the important parameter is the time at which the sodium temperature entering the pump increases by 10°C above the initial value. Automatic SCRAM of the reactor is based on this signal. Hence, the SCRAM demand time predicted by the three-dimensional model is 24.5 s. The same predicted by the multi-zone and single-zone models are 14 s and 30 s, respectively. Thus, as far as the pre-diction of SCRAM demand time is concerned, the multi-zone model pre-dictions are not conservative from the point of view of reactor safety. However, the prediction by the single-zone model is conservative, since it is delayed by 5.5 s, compared to the three-dimensional model predictions.

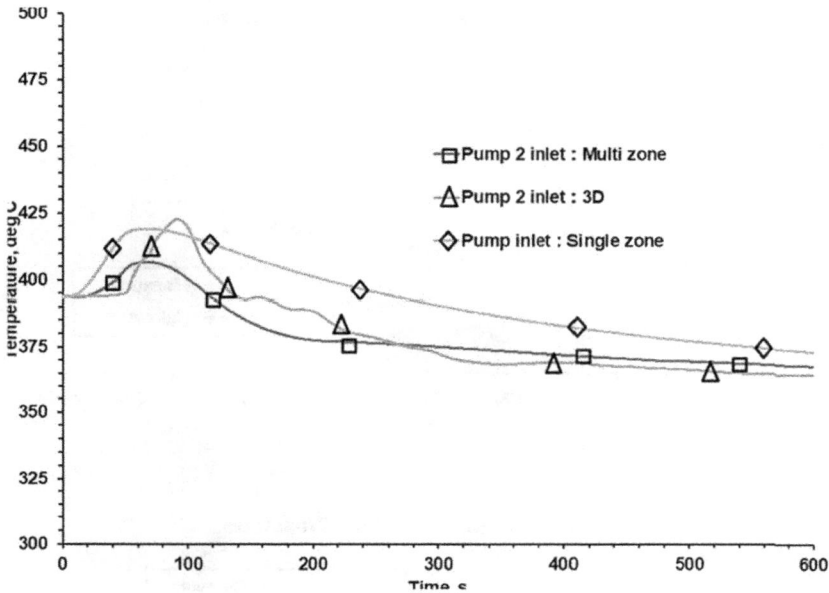

Figure 3.16 Predicted evolution of sodium temperature at inlet of pump 2 during one secondary pump trip.

Source: Natesan K., Development of mathematical models and investigation of FBR plant behaviour during coupled single and multi-dimensional approach, Doctoral Thesis, Homi Bhabha National Institute, Mumbai, India, 2020.

Thus, it can be concluded that as far as one-loop events are concerned, the predictions by the multi-zone model are neither conservative nor realistic. Hence, for system simulation and reactor safety, one-dimensional model would suffice for pool-type reactors. However, three-dimensional analysis may be necessary for assessing the detailed thermal loading on the cold pool structures.

Figure 3.17 shows the comparison between the predictions of a perfect mixing model with FBTR reactor measurements for the event of secondary pump trip in one loop. There is a fair comparison of predictions with measurements.

Equation 3.24 represents the one-dimensional model for a secondary loop design, and it is solved in the same manner as for the hot pool:

$$(MC)_m \frac{dT_m}{dt} = \sum_{i=1}^{2} Q_i C_{na} (T_i - T_m) \qquad (3.24)$$

Reactor Inlet junction - East loop

Reactor Inlet junction - West loop

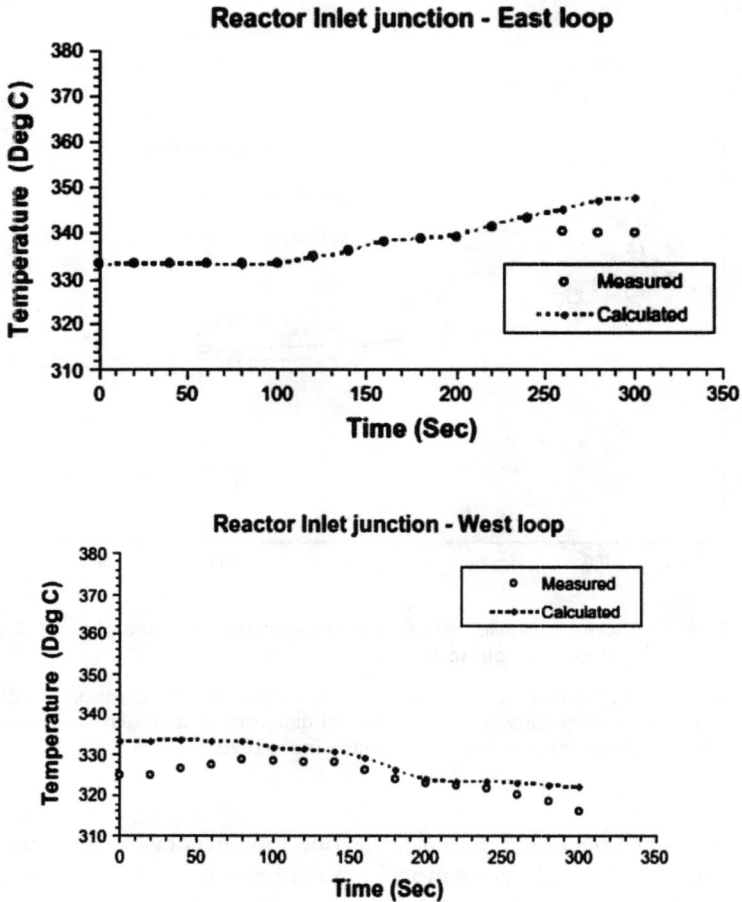

Figure 3.17 Reactor inlet temperature for one secondary pump trip event in FBTR.

Source: Vaidyanathan et.al, Dynamic Model of Fast Breeder Test Reactor. *Annals of Nuclear Energy* 37, 2010.

3.15 GRID PLATE

Grid plate is an important component of reactor assembly that supports the core subassemblies and serves as a plenum to distribute primary coolant flow through various subassemblies. The primary sodium pumped by the primary sodium pumps (PSP) is supplied to the grid plate through multiple pipes in case of pool type, unlike a loop type where the discharge pipes from the pumps join and a single pipe leads sodium to the grid plate. The grid plate contains several cylindrical sleeves in which core subassemblies are mounted. Primary sodium supplied to the grid plate through the primary pipes enters the subassembly through slots provided in the sleeves.

Since the grid plate in a pool-type reactor has multiple entries, studies were conducted for PFBR grid plate to assess the flow distribution through the different assemblies in case of a pipe rupture. Comparison of flows through different assemblies between three-dimensional and point models of the grid plate showed variations of +0.5% to −1.0%. Experiments carried out in a one-third scale model with air as the fluid confirmed the predictions (Sridhara Rao et al., 2001). Thus, the use of the point model of the grid plate is justified for system dynamic studies.

3.16 HEAT TRANSFER CORRELATIONS FOR FUEL ROD BUNDLE

Heat transfer correlations for steady-state turbulent flow in channels or rod bundles are of the form

$$Nu = C_1 + C_2 (\Psi \, Pe)^{C_3}$$

where Nu is Nusselt number, C_1, C_2, C_3 are constants, ψ is the average effective value of the ratio of eddy diffusivity of heat to that of momentum, and Pe is Peclet number (Reynolds number * Prandtl number). Dwyer (1976) has given the constants for a range of conditions. For liquid sodium flow parallel to rod bundles, Kazimi and Carelli (1976) have compiled available experimental data and recommend the following correlation:

$$Nu = 4.0 + 0.33 \left(\frac{P}{D}\right)^{3.8} \left(\frac{Pe}{100}\right)^{0.86} + 0.16 \left(\frac{P}{D}\right)^{5}$$

for $1.12 \le \dfrac{P}{D} \le 1.3; 10 \le Pe \le 5000$

$$Nu = \left[-16.15 + 24.96 \frac{P}{D} - 8.55 \left(\frac{P}{D}\right)^{2}\right] Pe^{0.3}$$

for $1.05 \le \dfrac{P}{D} \le 1.12; 150 \le Pe \le 1000$

$$Nu = 4.496 \left[-16.15 + 24.96 \frac{P}{D} - 8.55 \left(\frac{P}{D}\right)^{2}\right]$$

Table 3.3 Free Convection in Liquid Metals

Configuration	Correlation	Range
Free convection in liqu d sodium to a ccld horizontal plate	$Nu_D = 0.0785\ (Ra)^{0.32}$ $Nu_D = \dfrac{hD}{k}$, D = diameter of horizontal plate Ra = Rayleigh number based on D	Turbulent regime $5 \times 10^6 \leq Ra \leq 4 \times 10^7$
Free convection to a heated vertical plate	$Nu_x = 0.3(Ra_x)^{0.3}$ X = plate height Ra = Rayleigh number based on x	Laminar regime $Ra_x < 10^6$
Free convection from inside wall of a vertical vessel	$Nu_x = 0.16\ ((Ra_x)(Pr_x))^{0.3}$ X = total height of cylindrical wall r = radius of vessel	Both laminar and turbulent regimes
Free convection to a horizontal cylinder	$Nu_D = 0.53\ (Ra\ Pr)^{0.25}$ D = diameter of cylinder	Laminar regime $Re < 10^5$
Free convection across an enclosed liqu d–metal gap between plates	Vertical parallel plates: $Nu_D = 0.028\ (Ra_D)^{0.355}$ Horizontal parallel plates: $Nu_D = 0.043\ (Ra_D)^{0.33}$ D = distance between plates	Turbulent regime $4 \times 10^4 \leq Ra \leq 1 \times 10^8$
Free convection within an open-endзd channel	$Nu_D = 0.68\ (Ra)^{0.165}$ D = distance between plates	Creeping regime $10^{-3} \leq Ra \leq 25$

$$for\ 1.05 \leq \frac{P}{D} \leq 1.5,\ Pe \leq 150$$

where P is the pitch and D the diameter of the fuel pin.

Free convection heat transfer is possible during shutdown conditions of the reactor. Correlations for free convection in liquid metals (Tang et al., 1978) are compiled in Table 3.3.

ASSIGNMENT

1. Indicate the materials used in SFRs as fuel, coolant, and structures. While stainless steels are used in SFRs, they cannot be used in PWRs. Explain why.
2. Compare the prompt jump approximation method and the Runge-Kutta method of solution of the point kinetics equations for a typical PWR and SFR. Write a computer program and indicate the observations.
3. For a typical SFR fuel pin calculate the temperatures with different number of concentric rings. Is the statement made in this chapter regarding adequacy of two concentric rings justified?

REFERENCES

Aboanber S. and Nahla A. (2002), Generalization of the analytical inversion method for the solution of the point kinetics equations. *Journal of Physics A: Mathematical and General* 35:3245.

APDA-233, (1966), Report on the fuel melting incident in the Enrico fermi atomic power plant on October 5, doi:10.2172/4766757, https://www.osti.gov/servlets/purl/4766757

Chao Y.-A. and Attard A. (1985). A Resolution of the Stiffness Problem of Reactor Kinetics. *Nuclear Science and Engineering* 90(1):40–46.

Charagi S. and Thangasamy S. (1976), *A Comparative Study of Reactor Kinetics Solution Methods*, Power Plant Dynamics and Control Symposium, BARC, Mumbai.

Dwyer G.E. (1976), *Liquid Metal Heat Transfer*, Sodium NaK Engineering Handbook, Gordon & Breach, Science Publishers Inc., New York, Vol. 2.

Filip Gottfridsson, (2011), Simulation of Reactor Transient and Design Criteria of Sodium cooled Fast Reactors, Uppasala Universitat, UTH-enheten, Februari, https://www.diva-portal.org/smash/get/diva2:402326/FULLTEXT01.pdf

Gourdon J. (1979), "The detection of anamolies in Fast Reactor Cores", Inte. Top. Meeting on Fast Reactor Safety, Seattle.

Govindarajan S. and Vaidyanathan G. (1985), FACTT-A computer code for calculation of fuel and clad temperatures under transient conditions. In: IAEA-Proceedings IWGFR Specialists' Meeting on LMFBR Fuel Rod Behaviour Under Operational Transients, Kalpakkam, India, 1985.

Greenspan, H., Kelber, C.N. and Okrent, D. Eds. (1968), *Computing Methods in Reactor Physics*, Gordon and Breach Science Publishers, Inc., New York, NY, Chapter 6, pp. 444–506.

Hummel H. and Okrent D. (1970), "Reactivity coefficients in large Fast Power reactors", American Nuclear Society.

J.A.W. da Norbrega, (1971), "A New Solution of the Point Kinetics Equations," *Nuclear Sci. Engg.* 46, 366–375.

KAERI-TR-1105/98, (1998), Reactivity Feedback Models For SSC-K.

Kaganove J.J. (1960), "Numerical solution of one group space dependent Reactor Kinetics Equation", ANL-6132, Argonne National Lab. USA.

Kazimi M.S. and Carelli M.D. (1976), "Heat Transfer correlations for Analysis of CRBRP Assemblies", CRBRP-ARD-0034.

Laureau A., Lederer Y., Krakovich A., Buiron J. and Fontaine B. (2017). Transient coupled neutronics-thermal hydraulics study of ULOF accidents in sodium fast reactors using spatial kinetics, Proceedings of ICAPP 2017, April 24–28, 2017 – Fukui and Kyoto (Japan).

Lavarenne et al., (2019), A 2-D Correlation to evaluate fuel-cladding gap thermal conductance in mixed oxide fuel elements for sodium-cooled fast reactors, International Nuclear Fuel Cycle Conference (GLOBAL), Seattle, WA, USA, 22–26 September 2019 pp. 873–880.

Leonoapiders and John Seinfield N. (1971), *"Numerical Solution of ODEs"*, Academic Press, New York.

Maurin. (1982), "Creys Malville subassembly accidental Condition", Int. Top. Meeting on Fasty Reactor Safety, Lyon.

Natesan K. (2020), Development of mathematical models and investigation of FBR plant behaviour during coupled single and multi-dimensional approach, Doctoral Thesis, Homi Bhabha National Institute, Mumbai, India, http://www.hbni.ac.in/phdthesis/engg/ENGG02201304010.pdf

Neimark L., Lambert J.D.B., Murphy W.T. and Ringro C.W. (1972), "Performance of mixed oxide fuel element at 11 at% Burnup", *Nucl. Tech.*, Vol. 16, p. 75.

Raj B., Chellapandi P., and Vasudeva Roo P., (2015). *Sodium Fast Reactors with Closed Fuel Cycle*, CRC Press, 2015.

Rose P.F. and Burrows (1976), "ENDF/B Fission Product Decay Heat", Brookhaven National Lab., BNL-NCF-50545.

Sanchez, J. (1989), "On the Numerical Solution of the Point Kinetics Equations by Generalized Runge-Kutta Methods," *Nucl. Eng. Des.*, Vol. 103, pp. 94–99.

Sridhara Rao M.V., Padmakumar, G., Vaidyanathan, G., Prabhakar R., Ghosh D., Govindarajan S. and Kale R.D. (2001), Flow Distribution in the Grid Plate of Prototype Fast Breeder Reactor at different operating conditions, 28th National Conference on Fluid Mechanics and Fluid Power. 13–15 Dec 2001. Chandigarh.

Srinivasan G., Suresh Kumar K.V., Rajendran B. and Ramalingam P.V. (2006), The Fast Breeder Test Reactor—Design and operating experiences, *Nucl. Eng. Des*, Vol. 236, pp. 796–811; https://doi.org/10.1016/j.nucengdes.2005.09.024

Tang Y.S., Coffield R.D. and Markley R.A. (1978), "*Thermal Analysis of Liquid Metal Fast Breeder Reactors*", American Nuclear Society, La Grange Park, IL.

Vaidyanathan G., Kasinathan N. and Velusamy K. (2010), Dynamic Model of Fast Breeder Test Reactor, *Ann. Nucl. Energy*, Vol. 37. http://dx.doi.org/10.1016/j.anucene.2010.01.013

Vaidyanathan G. (2013), *Nuclear Reactor Engineering*, S. Chand Publishers, Delhi, India.

Yang J.W., Agrawal A.K., (1976), "An Analytical Model for Transient Fluid Mixing in Upper Plenum of an LMFBR," *Proceeding of the International Meeting on Fast Reactor Safety and Related Physics*, CONF-761001, 1448–1456, Chicago.

Chapter 4

IHX Thermal Model

4.1 INTRODUCTION

The intermediate heat exchanger (IHX) transfers the heat from the radio-active primary sodium to the nonradioactive secondary sodium. The IHX and complete secondary sodium system serve as a physical barrier between the radioactive sodium and the tertiary steam/water system. IHX used in the loop- and pool-type SFR concept is shown in Figures 4.1 and 4.2. In the loop type, primary sodium enters through a side nozzle on the shell, rises, and enters the tube bundle region through windows on the inner shell (Mochizuki and Takano, 2009). Knowledge of the crossflow velocities at the inlet are important in deciding the heat transfer area, tube bundle support locations, and their spacing for protecting the tube bundle from flow-induced vibration. Secondary sodium enters from the top of IHX through a central downcomer into a Toro spherical header, from where it gets distributed to different tubes. The pressure of secondary sodium is higher than that of primary one, so that in case of a leak, only nonradioactive sodium will flow into reactor sodium, avoiding any radioactivity in the secondary sodium loop. The downcomer is surrounded by another concentric shell with argon in the gap, to avoid heat transfer between downcoming secondary sodium and primary sodium. In the pool-type IHX, the difference is that primary sodium enters the IHX circumferentially from the hot pool (Figure 4.2) (Gajapathy, 2008). Here also the velocities of sodium entering the IHX influence the choice of supports for the tube bundle. In this chapter we present a description of IHX, details of its thermal modeling, and various solution techniques for realistic assessment of steady and transient temperatures.

4.2 EXPERIENCE IN PHENIX

The IHX in nearly all the reactors have had trouble-free operation except in the case of the PHENIX reactor. In this reactor in France, failure of a weld joint on the secondary outlet collector occurred due to differential

DOI: 10.1201/9781003283188-4

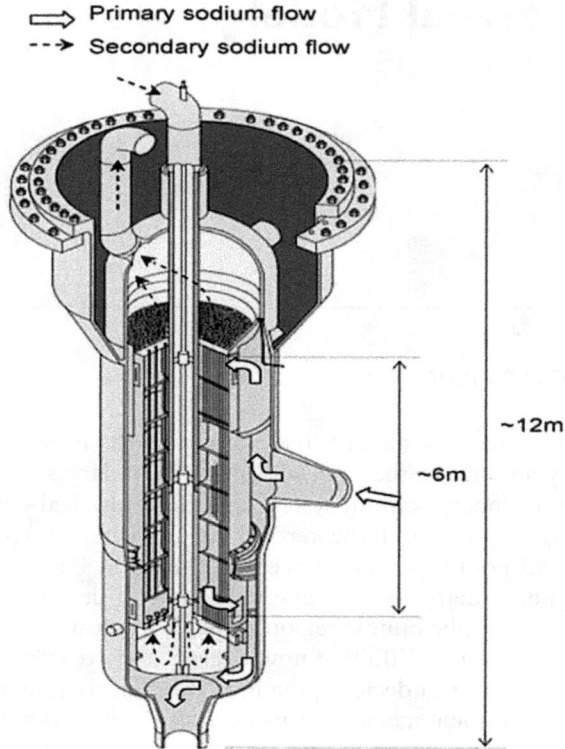

Figure 4.1 IHX for loop-type SFR. (Mochizuki, H. & Takano, M., Heat Transfer in
Heat Exchangers of Sodium Cooled Fast Reactor Systems. *Nuclear
Engineering and Design* **239**, 2009.)

expansion between the central downcomer tube and the outer shell (Conte
et al., 1977). This was traced to the thermal gradients because of cross-flow
of primary sodium at the IHX inlet. As the primary flow travels from outer
row of tubes to the inner row of tubes at the IHX inlet, the outermost row
sees sodium at the higher temperature (~530°C) while the inner row sees
sodium at lower temperatures (~480°C). Consequently, the heat transferred
to the outer rows of tubes is more than that to the inner row of tubes.

It was seen that a difference of ~50°C existed between the innermost and
outermost rows. This correspondingly was felt on the secondary side, and
due to inadequate mixing in the secondary outlet header, the structural temperatures were different, leading to considerable differential expansion.
Modifications were made in the secondary sodium outlet header where
sodium exiting from the outer row of tubes with higher sodium temperatures was directed toward the inner rows, thereby mixing it effectively and
reducing the differential temperatures. This was made possible with a mixer

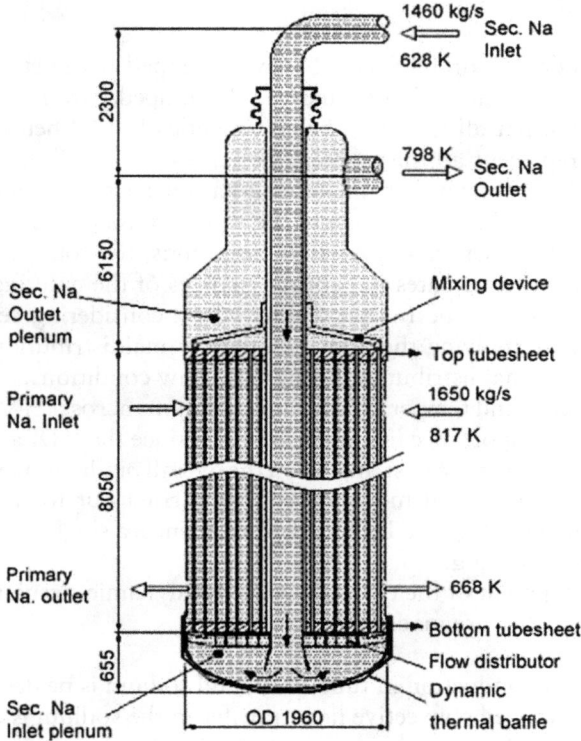

1460 kg/s
628 K — Sec. Na Inlet

798 K — Sec. Na Outlet

Sec. Na Outlet plenum

Mixing device

Primary Na. Inlet

Top tubesheet

1650 kg/s
817 K

Primary Na. outlet

668 K

Bottom tubesheet

Flow distributor

Dynamic thermal baffle

Sec. Na Inlet plenum

OD 1960

2300
6150
8050
655

Figure 4.2 IHX for pool-type SFR. (Gajapathy R., Thermal Hydraulic Investigations of Intermediate Heat Exchanger in a Pool-Type Fast Breeder Reactor. *Nuclear Engineering and Design* 238(7), 2008.)

at the secondary sodium outlet as shown in Figure 4.2. Subsequently, no further failures have taken place.

A study was carried out for PFBR to evolve a secondary sodium flow zoning concept like the core flow zoning. A suitable flow distribution device at the inlet to IHX tubes to ensure more flow to the outer rows and lesser flow to the inner rows was designed (Padmakumar et al., 2003). The sodium flow in the inner rows is reduced with the distributor and in the outer rows is increased. The impact of this on the secondary sodium temperature distribution was studied using 3D computational fluid dynamics by Suyambazhahan et al. (2014). It was found that the temperature variation for the secondary sodium at the exit, which was about 50°C when no flow distributor is present, reduced to about 25°C with the distributor plate. However, to bring down the radial temperature variation further, a mixing device has been introduced downstream of the tube bundle (apart from the flow distributor), as shown in Figure 4.2. Hence, it appears better to have a mixer at the secondary sodium tube outlet instead of a flow distribution device.

4.3 THERMAL MODEL

Thermal simulation of IHX can be by a lumped-parameter single-tube model or a detailed multi-channel model. The lumped-parameter single-tube model assumes that all the tubes behave identically, and hence an average tube should represent all the tubes. Experiences in modeling by Aburomia et al. (1975) have shown that the lumped-parameter single-tube and multi-channel models yield approximately the same transient results except for very low flows. Under these low-flow conditions, the total heat stored in the IHX elements attenuates the abrupt changes of the entering fluids. The single-tube model can be further improved by considering several parallel channels and applying the appropriate flow maldistribution factors, to account for flow maldistribution during low-flow conditions.

The radial flow and temperature maldistributions across IHX tube bundle are not of great importance in system analysis, since the IHX average outlet temperatures have the greatest impact on the overall prediction. Nevertheless, detailed multidimensional modeling can be carried out for assessment of detailed thermal loadings on IHX structures. Hence, a single-tube model can be used in the modeling.

In the development of the thermal model for dynamic simulation, the following assumptions are made:

- Axial heat conduction in tube, shell, and sodium is neglected.
- Fully developed convective heat transfer in the sodium is considered.
- The shell and downcomer are perfectly insulated.

With the above features, the energy balance over a differential length of the IHX gives rise to a set of coupled partial differential equations as given below.

$$\left(\rho V'C\right)_P \frac{\partial T_P}{\partial t} = -\left(WC\right)_P \frac{\partial T_P}{\partial Z} + \left(UA'\right)_P \left(T_T - T_P\right) + \left(UA'\right)_V \left(T_V - T_P\right) \quad (4.1)$$

$$\left(\rho V'C\right)_V \frac{\partial T_V}{\partial T} = \left(UA'\right)_V \left(T_P - T_V\right) \quad (4.2)$$

$$\left(\rho V'C\right)_T \frac{\partial T_T}{\partial t} = \left(UA'\right)_P \left(T_P - T_T\right) + \left(UA'\right)_S \left(T_S - T_T\right) \quad (4.3)$$

$$\left(\rho V'C\right)_S \frac{\partial T_S}{\partial t} = \left(WC\right)_S \frac{\partial T_S}{\partial z} + \left(UA'\right)_S \left(T_T - T_S\right) \quad (4.4)$$

where T_p, T_T, T_v, and T_s refer to temperatures of primary sodium, tube, shell, and secondary sodium, respectively; $(\rho V'C)$ is the mass capacity per unit length; W the mass flow rate; C the specific heat of sodium; and h the heat conductance.

4.4 SOLUTION TECHNIQUES

A brief discussion on the various methods is presented below.

4.4.1 Nodal Heat Balance Scheme

This scheme was suggested during the IHX analysis for FFTF (Gunby, 1970). In this approximation the metal wall nodes are placed in the mid-plane between the corresponding coolant nodes as shown in Figure 4.3, giving rise to a staggered nodal arrangement.

The coolant energy equation can be simplified by representing the convective term in a backward difference approximation as given below,

$$\frac{\partial T}{\partial z} \cong \frac{T_J - T_{J-1}}{\Delta z} + 0(\Delta z) \tag{4.5}$$

The spatial increment Δz must be sufficiently small to ensure that higher-order terms are negligible. Therefore, the integrated form of the partial differential equations reduces to the following ordinary differential equations:

$$\rho_J V_J \frac{dh_J}{dt} = W\left(h_{J-1} - h_J\right) + q_J \tag{4.6}$$

where V_j is the volume of cell, h_j is the enthalpy of fluid, and q_j the heat transfer rate.

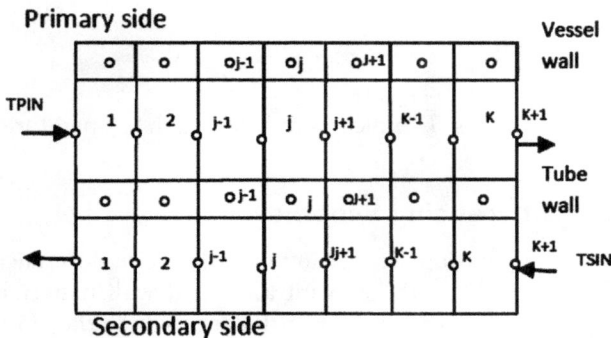

Figure 4.3 Lumped thermal model—staggered node.

The following gives the complete set of nodal heat balance equations of IHX.

Shell-side sodium

$$\left[\rho_S V_S \frac{dh_S}{dt}\right]_J = W_S\left(h_{S_{J-1}} - h_{S_J}\right) - q_{1_J} - q_{2_J} \tag{4.7}$$

Tube-side sodium

$$\left[\rho_t V_t \frac{dh_t}{dt}\right]_J = W_t\left(h_{t_{J-1}} - h_{t_J}\right) + q_{3_J} \tag{4.8}$$

Vessel (shell) structure

$$\left[\rho_V V_V C_V \frac{d\bar{T}_V}{dt}\right]_J = q_{1_J} \tag{4.9}$$

Tube-wall structure

$$\left[\rho_W V_W C_W \frac{d\bar{T}_W}{dt}\right]_J = q_{2_J} - q_{3_J} \tag{4.10}$$

$$q_{1_J} = U_1'\left(\bar{T}_S - \bar{T}_V\right)_J \tag{4.11}$$

$$q_{2_J} = U_2'\left(\bar{T}_S - T_W\right)_J \tag{4.12}$$

$$q_{3_J} = U_3'\left(\bar{T}_W - \bar{T}_t\right)_J \tag{4.13}$$

In the above equations $\bar{T}$ indicates average mesh temperatures.

4.4.2 Finite Differencing Scheme

The numerical computation scheme involves the finite differencing scheme for the dT/dZ terms. Here the coolant and solid wall material nodes are placed in parallel, unlike the nodal heat balance where they are staggered. The convective term in the energy balance can be represented by the following approximations.

Two-Point Backward Difference

$$\frac{dT}{\partial z} \cong \frac{T_J - T_{J-1}}{\Delta z}$$

Three-point backward difference

$$\frac{\partial T}{\partial z} \cong \frac{3T_J - 4T_{J-1} + T_{j-2}}{2\Delta z}$$

Central difference

$$\frac{\partial T}{\partial z} = \frac{T_{J+1} - T_{J-1}}{2\Delta z}$$

Mixed difference

$$\frac{\partial T}{\partial z} = \frac{T_{J+1} - T_{J-1}}{2\Delta z} \text{ for an even number of nodes}$$

$$\frac{\partial T}{\partial z} = \frac{3T_J - 4T_{J-1} + T_{j-2}}{2\Delta z} \text{ for an odd number of nodes}$$

4.5 CHOICE OF NUMERICAL SCHEME

A comparative study of the different solution techniques was made by Gunby (1970) in his report on IHX simulation for the fast flux test facility (FFTF). He concluded that the NHB and the mixed difference schemes are the best. He warns, however, against their use for severe transients involving highly unbalanced flow conditions.

It is important to note that the assumption of linear temperature distribution within a node may fail under very abrupt transient conditions, especially under unbalanced-flow conditions when the ratio of primary to secondary flows is far from 1. Under such conditions with varying transit times of primary and secondary sodium, the number of meshes and mesh size would not be realistic. In the following section the reasons for such behavior and some improvements to the nodal heat balance are presented.

4.5.1 Nodal Heat Balance for Unbalanced Flows

Nodal heat balance scheme with four nodes was applied to a steady-state solution of the IHX for FBTR under unbalanced flow conditions of W_s (shell-side sodium flow) and W_t (tube-side sodium flow) of 145 Kg/s and 20 Kg/s, respectively (Kasinathan et al., 1989). The results of the solution are given in Figure 4.4. Instead of an expected monotonically increasing or

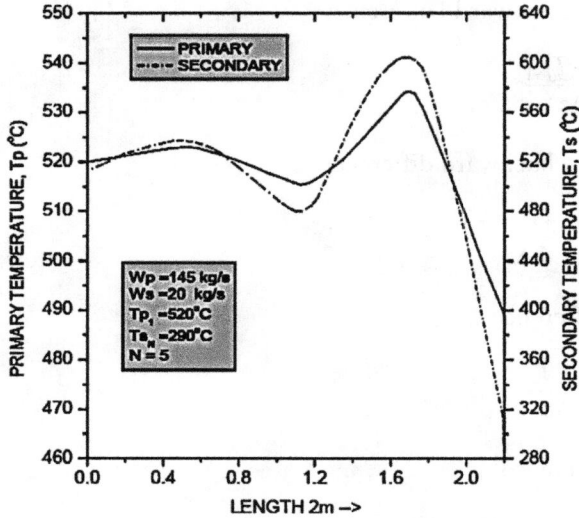

Figure 4.4 Steady-state temperature distribution—unbalanced flows. (Kasinathan N., Rajakumar A., Vaidyanathan G., Thermal Modelling of an IHX in a Liquid Metal Fast Breeder Reactor. *J. Indian Chemical Engineer* 31, 1989.)

decreasing temperatures, we see an oscillatory profile. A transient analysis for a secondary flow coast-down from a steady-state value of 97 Kg/s to 4.55 Kg/s in 5 s was conducted with the same methodology. The results, given in Figure 4.5, show an unrealistic dip in the outlet temperature of secondary sodium.

During low secondary flow conditions, the power removed by secondary would be small and all heat transfer may happen within one nodal length from the cold end. There would be no heat transfer in the further nodes, as the primary and secondary are already very close. The primary side temperatures would be at their inlet value for most of the length from the hot end of IHX, and the secondary fluid remains at its outlet temperature for most part of the IHX length. Under these conditions, the nodal bulk temperatures should preferably be an appropriate weighted mean of the adjacent grid point temperatures, instead of the arithmetic mean used.

4.5.2 Modified Nodal Heat Balance Scheme (MNHB)

Based on the above, Kasinathan et al. (1989) proposed a modified nodal heat balance scheme in which the nodal bulk temperatures are taken as weighted means of adjacent temperatures as given below:

$$\bar{T}_{pj} = F_p\, T_{pj} + \left(1 - F_p\right) T_{pj} - 1$$

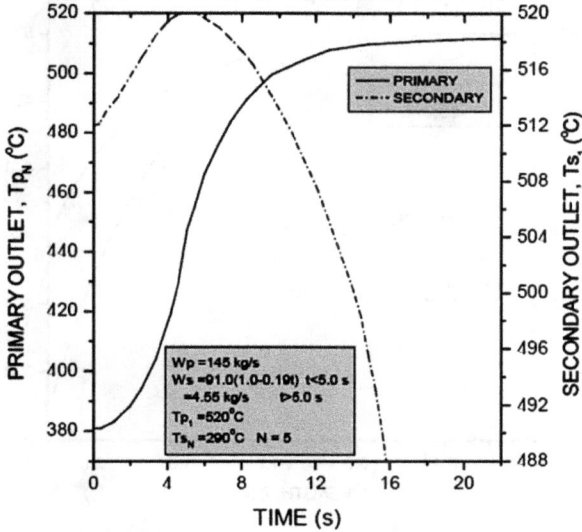

Figure 4.5 Transient response—unbalanced flows. Kasinathan N., Rajakumar A., Vaidyanathan G., Thermal Modelling of an IHX in a Liquid Metal Fast Breeder Reactor. *J. Indian Chemical Engineer* 31, 1989.)

$$\bar{T}_{sj} = F_s T_{sj} + \left(1 - F_s\right)T_{sj} - 1$$

In the above, F_p and F_s are the weighting factors for primary and secondary, respectively. For the case of the unbalanced flow analyzed for steady state, the value of F_p is taken as zero and F_s as 1.0. The analysis results are given in Figure 4.6. Spatial oscillations are eliminated and realistic solution obtained. A question arises how to take these factors for varying ratios of primary and secondary flows.

For unbalanced flows, as in the case of decreasing secondary sodium flow with primary sodium flow remaining constant, steep and highly non-linear temperature profiles are expected at the cold end of IHX. In any transient from such a balanced flow condition, the values of F_p and F_s are to be changed in a continuous manner depending on the temperature gradients prevailing in the different nodes. For the case of secondary flow decay, the weighting factors of F_p and F_s are computed based on the ratio of temperature gradients in the last, but one node and the last node (α) as given below:

$$\alpha_p = \left(T_{p_{j-2}} - T_{p_{j-1}}\right) / \left(T_{p_{j-1}} - T_{p_j}\right)$$

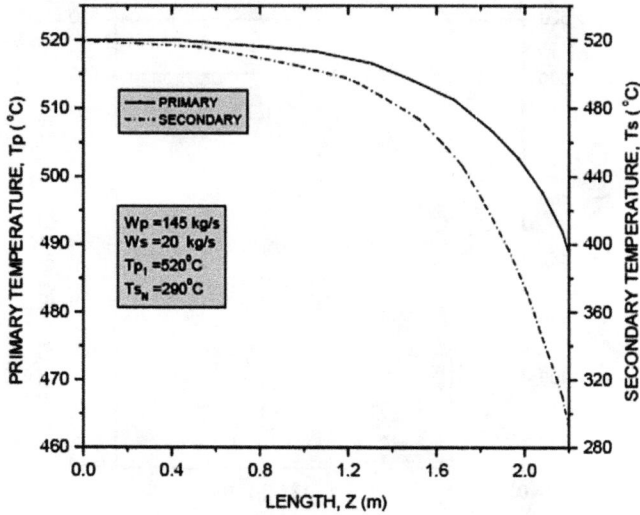

Figure 4.6 Steady-state response unbalanced flows—MNHB. (Kasinathan N., Rajakumar A., Vaidyanathan G., Thermal Modelling of an IHX in a Liquid Metal Fast Breeder Reactor. *J. Indian Chemical Engineer* 31, 1989.)

$$\alpha_s = \frac{\left(T_{s_{j-2}} - T_{s_{j-1}}\right)}{\left(T_{s_{j-1}} - T_{s_j}\right)}$$

and

$$F_p = 0.5\alpha_p \quad \text{and} \quad F_s = \left(1 - 0.5\alpha_s\right)$$

With the above formulations incorporated into the earlier nodal heat balance scheme, the solutions were obtained for the secondary flow reduction from 91 kg/s to 4.55 kg/s in 5 s. Results given in Figure 4.7 show a monotonous temperature profile. Thus, the unrealistic response obtained earlier with the nodal heat balance scheme has been overcome by the modified scheme. Hence the modified nodal heat balance (MNHB) is recommended for a plant dynamic simulation.

Agarwal et al. (2010) compared the MNHB scheme with various finite difference and finite volume techniques and arrived at the conclusion that the MNHB scheme was the best in terms of accuracy and computation, and this is recommended for the future. With the high heat transfer coefficients of sodium, the film drops are small, and hence further simplification can be made by clubbing half the capacity of the tube wall and full shell along with primary sodium and clubbing the remaining half capacity of the tube wall with secondary sodium. The equation can then be solved by the MNHB scheme.

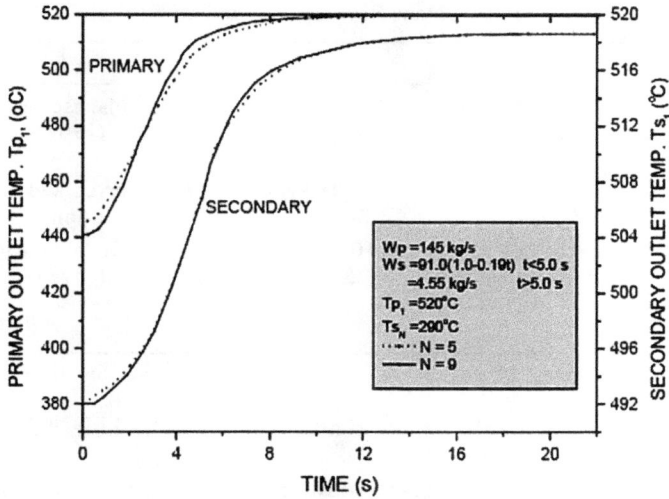

Figure 4.7 Transient response unbalanced flows—MNHB. (Kasinathan N., Rajakumar A.,Vaidyanathan G., Thermal Modelling of an IHX in a Liquid Metal Fast Breeder Reactor. *J. Indian Chemical Engineer* 31, 1989.)

4.6 HEAT TRANSFER CORRELATIONS

The convection heat transfer coefficient represented by the Nusselt number depends on the flow Reynolds's number, tube wall roughness, fluid Prandtl number, L/D of the tubes, etc. For fully developed turbulent flow in long tubes, a number of correlations have been suggested.

Table 4.1 gives widely used Nusselt number correlation for round tubes, with uniform heat flux at the wall. Under fully developed laminar flow conditions through tubes with uniform wall heat flux, Nu is 4.36. For shell-side

Table 4.1 Turbulent flow inside round tubes

Nusselt number correlations		Reference
$Nu = 7. + 0.025\,(\bar{\psi}\,Re.Pr)^{0.8}$	$\bar{\psi} = 1$	Lyon (1951)
	$\bar{\psi} = 1 - \dfrac{1.84}{Pr\left(\varepsilon_m/v\right)_{max}^{1.4}}$	Dwyer (1963, 1976)
$Nu = 6. + 0.025\,(\bar{\psi}\,Re.Pr)^{0.8}$	$\bar{\psi} = \dfrac{0.014\left(1-e^{-71.8\beta}\right)}{\beta}$	Aoki (1963)
	$\beta = Re^{-0.45}Pr^{-0.2}$	
$Nu = 5. + 0.025\,(\bar{\psi}\,Re.Pr)^{0.8}$	$\bar{\psi} = 1$	Subbotin (1964)

Table 4.2 Nu number for turbulent flow through unbaffled tube banks

Nusselt number correlation	Range	Reference
$Nu = 6.66 + 3.126 \, (P/D) +$ $1.184 \, (P/D)^2 + 0.0155$ $(\bar{\psi} \, Re.Pr)^{0.86}$	$70 \leq Re.Pr \leq 10^4$ $1.3 \leq P/D \leq 3.0$	Maresca and Dwyer (1964)
$Nu = 7.0 + 3.8 \, (P/D)^{1.52} + 0.027$ $(P/D)^{0.27} \, (\bar{\psi} \, Re.Pr)^{0.8}$	$0 < Re.Pr \leq 10^5$ $1.3 \leq P/D \leq 10$	Friedland and Bonilla (1961)
$Nu = \alpha + \beta \, (Re. \, Pr)^{\gamma}$ $\alpha = 0.25 + 6.2 \, (P/D)$ $\beta = -0.007 + 0.032 \, (P/D)$ $\gamma = 0.8 - 0.024 \, (P/D)$	$110 \leq Re.Pr \leq 4300$ $1.25 \leq P/D \leq 1.95$	Graeber and Rieger (1973)

heat transfer through tube banks that are unbaffled, Table 4.2 gives the commonly used correlations.

For heat transfer from primary sodium to secondary sodium, both the cross-flow and axial flow of primary sodium over the bundle are to be considered in the estimation of effective heat transfer coefficient.

The following heat transfer correlations used for the IHX design of Super PHENIX 1 reactor (Marcellin and Guidez, 1984) are given below.

$$Tube : Nu = 4.82 + 0.0185 \, Pe^{0.827}$$

$$Shell : 6.0 + 0.006 \, Pe$$

$$Cross \; flow : Nu = 4.03 + 0.228 \, Pe^{0.67}$$

where Nu and Pe are Nusselt and Peclet numbers, respectively. If h_c is the cross-flow heat transfer coefficient and h_a is the axial flow heat transfer coefficient from the above correlations, then the effective shell-side heat transfer coefficient, h_p, given by

$$\frac{1}{h_p^2} = \frac{\left(\left(U^2 / h_c^2 \right) + \left(V^2 / h_a^2 \right) \right)}{\left(U^2 + V^2 \right)}$$

where U and V are cross-flow and axial flow velocities, respectively.

4.7 VALIDATION

The model has been validated based on the commissioning tests in the FBTR (Vaidyanathan et al., 2010). The event of trip of one primary pump was analyzed by using the above model in conjunction with other plant models

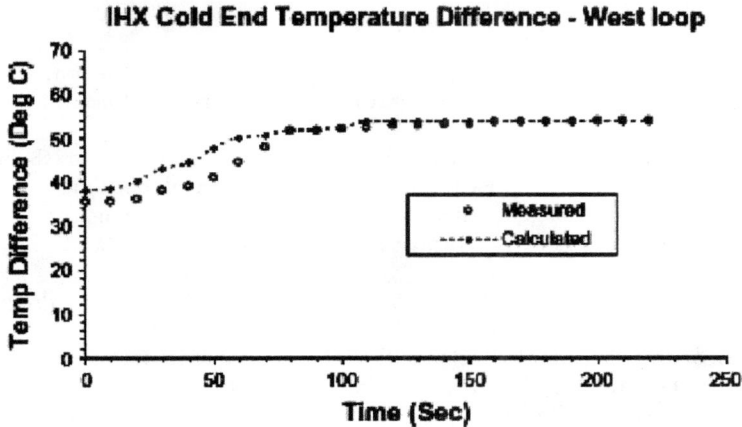

IHX Cold End Temperature Difference - West loop

Figure 4.8 IHX cold end temperature difference in the west loop. Vaidyanathan G., Kasinathan N., and Velusamy K., Dynamic Model of Fast Breeder Test Reactor. *Annals of Nuclear Energy* 37, 2010.)

in the DYNAM code. In this event, the primary pump in east loop is tripped. The reactor is tripped immediately based on the safety logic. Also, the water flow in the steam generator is reduced to a minimum value of 20% of the full flow to limit the cold thermal shocks on the components. The important effect is a rise in the temperature difference between the primary outlet and the secondary inlet in IHX of the unaffected west loop. This is essentially due to reduction in heat removal in that loop because of water flow reduction. It rises from 38°C at steady state to 57°C in 200 s and then falls as per DYNAM predictions (Figure. 4.8). There is a difference in the initial steady-state temperatures between predictions and plant measurements, as the actual plant initial power levels were slightly different than that used in the predictions. Nevertheless, the measured transient temperatures on the plant are very close to the predictions, as can be seen in Figure 4.8.

ASSIGNMENT

1. What is the basis of having secondary flow inside tubes and primary sodium on the shell side in an IHX?
2. Browse the literature on the IHX failure in the PHENIX reactor and describe in how many ways this issue could be resolved.
3. For a typical IHX solve the thermal equations with nodal heat balance method for unbalanced flows under steady and transient conditions and judge for yourself whether the conclusions drawn in this chapter are correct. You may need to write a program. Relevant data can be taken from literature indicated in references.

REFERENCES

Aburomia M.M., Cho S.M. and Sawyer S.W. (1975), "Thermal hydraulic design considerations for clinch river breeder reactor plant intermediate heat exchanger", ASME paper 75-WA/HT-101.

Agarwal, Sourabh, Revathy K., Banerjee I., Padma Kumar G., Anand Babu and Vaidyanathan G., (2010), "Experience on transient thermal modeling of sodium-to-sodium heat exchanger using FDM and FVM", *Proceedings of the 18th International Conference on Nuclear Engineering*, ICONE18, May 17-21, 2010, Xi'an, China.

Aoki S. (1963), "Current liquid metal heat transfer research in Japan", *Progress in Heat and Mass Transfer*, Vol. 7, p. 569.

Conte F., Sauvage M. and Roumelhac J.F. (1977), "The intermediate heat exchanger leak in PHENIX and their repair", *Proc. Int. Conf. Optimization of Sodium cooled fast reactors, BNES*.

Dwyer G.E. (1963), "Eddy transport in liquid metal heat transfer", *AIChE Journal*, Vol. 9, p. 2.

Dwyer G.E. (1976), "Liquid metal heat transfer", *Sodium NaK Engineering Handbook*, Gordon & Breach, Science Publishers Inc., New York, Vol. 2.

Friedland A.J. and Bonilla C.F. (1961), "Analytical study of Heat transfer rates for parallel flow of liquid metals through tube bundles-Part II", *AIChE Journal*, Vol. 7, p. 1.

Gajapathy R., (2008), "Thermal hydraulic investigations of intermediate heat exchanger in a pool-type fast breeder reactor", *Nuclear Engineering and Design*, Vol. 238, No. 7, pp. 1577–1591. https://doi.org/10.1016/j.nucengdes.2008.01.005

Graeber H. and Rieger M. (1973), "Experimental study of Heat Transfer to Liquid Metals flowing inline through tube bundles", *Progress in Heat and Mass Transfer*, Vol. 7, p. 151.

Gunby A.L. (1970), "IHX modeling for FFTF simulation", *BNWL-1367*.

Kasinathan N., Rajakumar A., Vaidyanathan G., (1989), "Thermal modelling of an IHX in a liquid metal fast breeder reactor", *J. Indian Chemical Engineer*, Vol. 31. pp. 45–49.

Lyon R.N. (1951), "Liquid metal heat transfer coefficient", *Chem. Eng. Prog.*, Vol. 47, p. 75.

Marcellin C. and Guidez J. (1984), "Hydrodynamic behaviour of intermediate heat exchangers in a pool-type fast breeder reactor", *Liquid Metal Engg. and Tech.*, BNES, London.

Maresca M.W. and Dwyer O.E. (1964), "Heat transfer to mercury flowing inline through a bundle of circular rods", *J. Heat Transfer*. May 1964, 86(2): 180–186, https://doi.org/10.1115/1.3687091

Mochizuki, Hiroyasu and Takano, Masahito. (2009), "Heat transfer in heat exchangers of sodium cooled fast reactor systems", *Nucl. Eng. Des.* Vol. 239, pp. 295–307. http://dx.doi.org/10.1016/j.nucengdes.2008.10.013

Padmakumar G., Sundaramoorthy T.R., Vaidyanathan G., Prabhakar R., Pai S.S. and Konnur M.S. (2003), "Flow distribution device for PFBR IHX", *Proceedings of 11th International Conference on Nuclear Engineering*, Tokyo, Japan.

Subbotin V.I. (1964)," Heat removal from reactor fuel elements cooled by liquid metals", *Proc. Third Int. Conf. on Peaceful uses of Atomic Energy, Conf-28/P3288*, Geneva.

Suyambazhahan S., Sarit K. Das, Velusamy K., and Sundararajan T. (2014), "A computational study of flow mal-distribution on the thermal hydraulic performance of an intermediate heat exchanger in LMFBR", *Journal of Nuclear Science and Technology*, Vol. 51, No. 6, pp. 845–857, http://dx.doi.org/10.1080/00223131.2 014.903812

Vaidyanathan G., Kasinathan N. and Velusamy K., (2010), "Dynamic model of Fast Breeder Test Reactor", *Annals of Nuclear Energy*, Vol. 37, pp. 450–462. https:// doi.org/10.1016/j.anucene.2010.01.013

Chapter 5

Thermal Model of Piping

5.1 INTRODUCTION

The coolant transport in the primary and secondary heat transport system is one of the most important effects, because the coolant spends reasonable time in the piping, especially the secondary sodium system. In an SFR, the IHX is located inside the reactor containment building, while the steam generators, secondary sodium pumps, and expansion and surge tanks are in the steam generator building. All these components are connected by sodium piping of long lengths. The piping runs are extensive with loops and bends. Hence, coolant transport in the piping should form an important part of the overall system simulation model. The piping can be subjected to thermal stresses induced by thermal mixing or thermal stratification. In case of mixing between hot and cold sodium, temperature fluctuations occur and can cause thermal fatigue on the pipe. In the case of improper mixing, especially at low sodium flows (thermal stratification), the hot and cold streams could flow parallel and cause circumferential stresses on the pipe. The thermal response at different components would depend on the accuracy of the piping thermal model. Hence, energy transport through piping needs attention.

The temperature distribution in the piping is needed for accurate assessment of buoyancy forces, which are important while assessing natural convection flow. The loops and bends in piping are provided to accommodate thermal expansion of the piping during operation, without exceeding acceptable stress limits. This needs the temperature distribution in the piping.

Generally, the temperature distribution is obtained by first converting the partial difference equations using finite differences with space derivative terms to obtain ordinary differential equations. Gerhardstein (1966) used central and backward differencing. Jennings (1969) compared several multi-node models, including the finite-difference method (backward and central difference) and the nodal heat balance method (applying energy balance over a nodal control volume), and concluded that nodal heat balance models were better. Gunby compared models under variable temperature and

DOI: 10.1201/9781003283188-5

flow transients (Gunby, 1970). A mixed-difference (alternating forward and three-point backward difference) model was also added to the comparisons. He concluded that backward and central differences were not good from the point of accuracy and response. The mixed difference was considered best, with the nodal heat balance a close second and hence a good alternative. In this chapter the different approaches to obtaining temperatures are presented and a suitable scheme recommended for simulation.

5.2 THERMAL MODEL

The characteristic features of the modeling are:

- one-dimensional flow, i.e., uniform velocity and temperature profiles normal to the flow direction
- incompressible, single-phase liquid
- single mass flow rate
- fully developed flow and heat transfer
- no axial heat conduction in the walls

For an idealized adiabatic flow in a pipe system, the transient temperature at the outlet of a pipe section is equivalent to the transient temperature at its inlet delayed by the transport time. This delay will vary continuously in the case of a pump trip or a coast-down. This is known as a time delay model. This is not realistic, as it does not account for turbulent mixing in the pipe, as well as heat storage in the pipe walls. Both these factors can have a major effect on the transient temperatures. An improvement over the time delay model is the mixing model. Here again the heat storage in pipe walls does not enter the calculations. Next is the consideration of coolant and wall, where heat storage in the wall is considered (Figure 5.1). The most sophisticated would be the use of coolant, wall, and insulation model with heat losses to the environment considered.

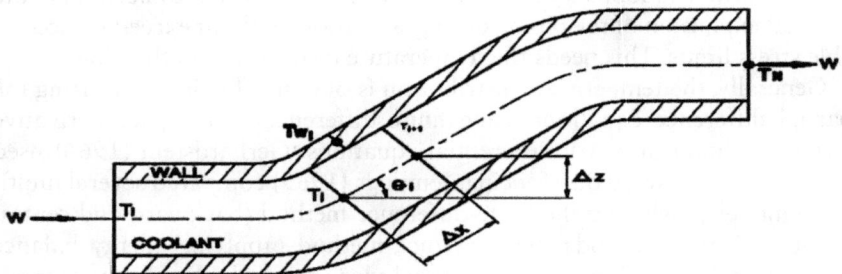

Figure 5.1 Piping schematic.

The governing equations for the last model are given below:
Sodium to pipe wall

$$\left(MC\right)_{Na}\frac{\partial T_{Na}}{\partial t}+QC_{Na}\frac{\partial T_{Na}}{\partial Z}=-h_1\left(T_{Na}-T_w\right) \tag{5.1}$$

Pipe wall to insulation

$$\left(MC\right)_w\frac{\partial T_w}{\partial t}=h_1\left(T_{Na}-T_W\right)-h_2\left(T_w-T_{in}\right) \tag{5.2}$$

Insulation to air

$$\left(MC\right)_{in}\frac{\partial T_{in}}{\partial t}=h_2\left(T_w-T_{in}\right)-h_3\left(T_{in}-T_a\right) \tag{5.3}$$

Where T is temperature, t is time, h is heat transfer coefficient, MC is mass capacity, while subscripts Na, w, and a, refer (respectively) to sodium, wall ambient, and insulation, and h_1, h_2, h_3 represent the heat transfer coefficient between (respectively) sodium and wall, wall and insulation, and insulation and ambient air.

The solution of these equations would yield the temperature distribution across piping. The heat transfer coefficients for sodium can be calculated based on the correlations given in Table 4.1 in Chapter 4. Sodium thermal conductivity being high, the overall heat transfer would be governed by the wall resistance, and hence uncertainty in sodium heat transfer coefficients will not impact the temperatures of the wall.

5.3 SOLUTION METHODS

The piping is divided into a suitable number of meshes (Figure 5.2), and for each mesh the differential equations are converted into algebraic form using the finite difference approximations as follows:

$$T=\left(T_{ij}+T_{ij-1}\right)*0.5$$

$$\frac{\partial T}{\partial t}=\frac{\left(T_{ij}+T_{ij-1}\right)}{\Delta j}$$

$$\partial T\,/\,\partial z=\left(\left(T_{ij}+T_{ij-1}\right)-\left(T_{ij}+T_{i-1j-1}\right)\right)/2\,/\Delta i$$

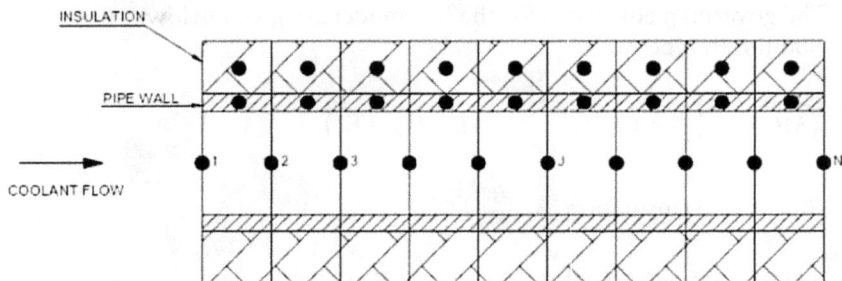

Figure 5.2 Staggered mesh piping thermal model.

Where i and j refer to spatial and time meshes.

The algebraic equations of sodium, wall, and insulation are solved by successive substitution. A slight variation of the above method is making the equations semi-implicit with the following substitutions:

$$\frac{\partial T}{\partial t} = \frac{\left(T_{ij} - T_{ij-1}\right)}{\Delta j}$$

$$\frac{\partial T}{\partial z} = \frac{\left(\left(T_{ij} - T_{i-1j}\right)\right)}{\Delta i}$$

Here the current time step values appear on both sides of the equation, unlike the previous method.

5.4 COMPARISON OF PIPING MODELS

The coolant-wall-insulation (CWI) model is taken as the reference model in which heat transfer through the coolant, wall, and insulation are considered as individual entities. In the next model the effect of insulation is ignored and coolant-wall with adiabatic wall boundary is assumed (CW). In other words, heat losses from piping are ignored. In the coolant mixing model (CMIX), again no heat loss is considered and perfect mixing is assumed. The transport delay model (TDM) is based solely on the time delay in transportation of coolant.

The sensitivity of the predicted results to the degree of axial modulization was examined by Madni and Agrawal (1980). A sample transient representing a severe temperature down-ramp of 127 K in 80 s (which may occur at a heat exchanger outlet) was applied to a pipe section (Figure 5.3) with flow decaying with a flow halving time of 5.5 s, which corresponds to that for a CRBRP plant. Figure 5.3(a) compares the predicted transient

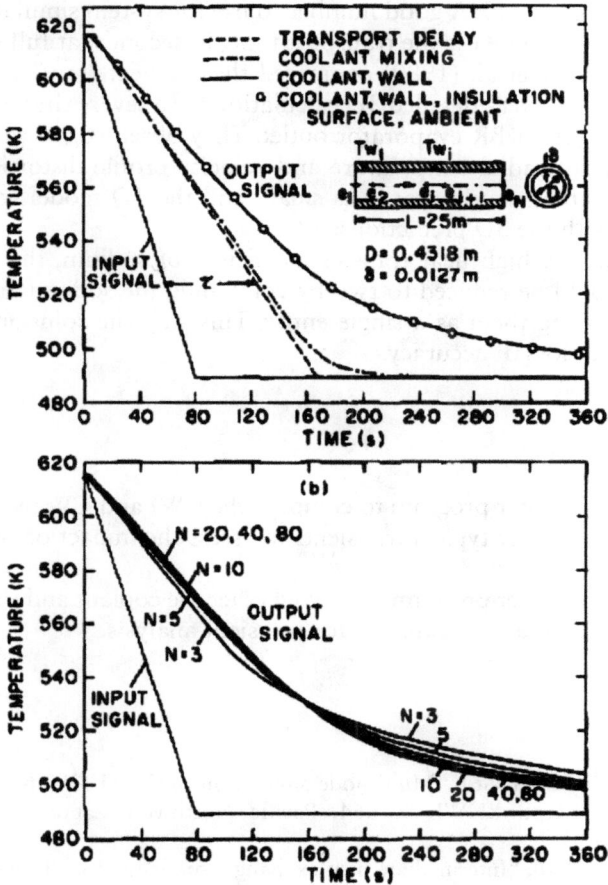

Figure 5.3 (a) Comparison of predicted temperatures using various models; (b) sensitivity of results to axial nodalization. (Imtiaz K. Madni and Ashok K. Agrawal, LMFBR System Analysis: Impact of Heat Transport System on Core Thermal-Hydraulics. *Nuclear Engineering and Design* 62, 1980.)

outlet temperature response of the CW model with CWI model that includes heat losses through insulation. Also included in the comparison are the transport delay model and a coolant mixing (adiabatic wall) model. Excellent agreement is observed between the CW model and the more detailed CWI model, the maximum deviation being less than 0.1 K. The insulation conductivity used was 0.065 W/mK. When this was doubled to 0.13 W/mK, the deviation was still less than 0.2 K. In contrast, the transport delay and mixing models are seen to grossly underpredict the outlet temperature. Figure 5.3(b) examines the sensitivity of the predicted output signal to axial nodalization. The maximum deviation between N = 3 and N = 5 is 4 K, decreasing to less than 0.5 K between N = 20 and N = 40, and converging for N > 40. A good number to use for system simulation would be N ~ 2(t+ 1), where t is the transport time, in seconds, at full flow.

Khatib-Rahbar et al. (1980) compared the 1D heat transfer model discussed above to a very detailed 3D calculation for a severe thermal transient occurring at the CRBR evaporator outlet. They observed for this case that the influence of radial temperature and velocity profile distortions on the axial temperature distribution was small, and the 1D model was in good agreement with the 3D predictions.

In view of the high heat transfer coefficient of sodium, the number of equations could be reduced to two by combining the sodium and the pipe wall and treating them as a single entity. This way the solution would be faster without loss of accuracy.

ASSIGNMENT

1. Write a computer program to compare the CWI and CW models of piping and test it for typical transients to assess the impact on temperature calculation.
2. Using a computer program, justify whether the coolant and the pipe wall can be treated as one capacity for transient analysis.

REFERENCES

Gerhardstein L.H., (1966), "Multi-node simulation of the FFTF intermediate heat exchanger", Rept. BNWL-CC-664, Pacific Northwest Laboratory, Richland, Wash., May.

Gunby A.L., (1970), "Intermediate heat exchanger modeling for FFTF simulation", Rept. BNWL-1367, Pacific Northwest Laboratory, Richland, Wash., May.

Jennings G.J., (1969), "Mathematical models of heat exchangers", Westinghouse Rept. FPC-113, Pittsburgh, PA.

Khatib-Rahbar M., Madni I.K. and Agrawal A.K., (1980), "Impact of multi-dimensional effects in LMFBR piping systems", *Proceedings of the Specialists Meeting on Decay Heat Removal and Natural Convection in FBRs, Brookhaven National Laboratory*, Feb. 28–29.

Madni, Imtiaz K. and Agrawal, Ashok K., (1980), "LMFBR system analysis: impact of heat transport system on core thermal-hydraulics", *Nuclear Engineering and Design*, Vol. 62, pp. 199–218.

Chapter 6

Sodium Pump

6.1 INTRODUCTION

The primary concern in the safety analysis of a SFR system is whether adequate cooling capability is provided to maintain the fuel element temperatures below specified values during off-normal and accident events that may occur in the heat transport system, e.g., loss of all pumping power. The behavior of the centrifugal pumps that circulate the reactor coolant becomes extremely important during such transients, and the ability to predict pump performance is necessary to understand and predict the interrelated hydraulic phenomena controlling loop and core flow rates.

Pumps used in SFRs are of centrifugal type or electromagnetic type. Sodium being a good electrical conductor makes it amenable for pumping through electromagnetic means. The advantage of electromagnetic pumps is that they are nonintrusive, and the electromagnet can be kept outside the pipe. Unfortunately, they have low efficiency of the order of ~20%. With developments in reliable mechanical seals, centrifugal sodium pumps with ~85% efficiency have been used in the main heat transport systems of nearly all the SFRs. Moreover, these pump drives can be provided with flywheel to limit the rate of flow reduction in case of a pump trip. Electromagnetic pumps have been used mostly in auxiliary circuits, like purification, fill and drain, etc. More recently, such pumps have been developed for larger flows. This chapter briefly details the construction of the pumps and describes the modeling of centrifugal pumps that are necessary for estimating the sodium flow in the primary and secondary circuits.

6.2 ELECTROMAGNETIC PUMPS

The electromagnetic (EM) pumps have no moving parts, which makes them maintenance free. Sodium is hermetically sealed in EM pumps, which eliminates the problem of sodium leakage. In EM pumps, an electric current is forced (via either conduction or induction) to flow through liquid metal. When the current carrying liquid metal is placed in the magnetic field, it

DOI: 10.1201/9781003283188-6

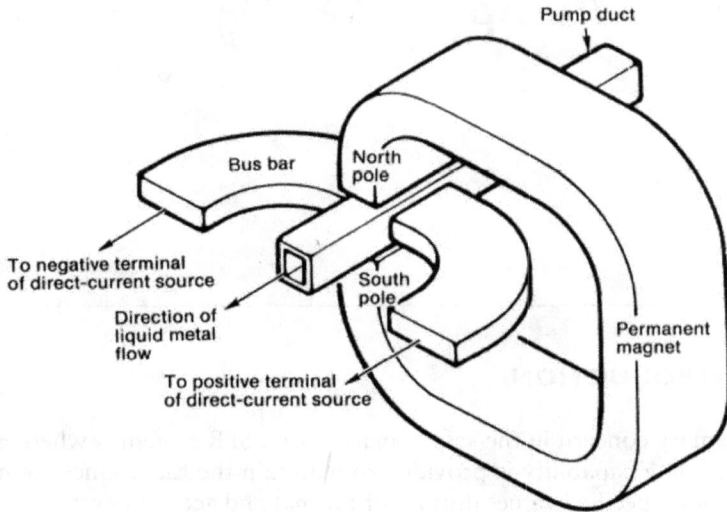

Figure 6.1 DC conduction pump.

experiences an electromagnetic body force (Lorentz force). The direction of force is decided by Fleming's left-hand rule. Force (F) = B I L, where B is the flux density, I is the current, and L is the length of the current-carrying portion of liquid metal. All the parameters mentioned above are at 90 degrees to each other. Figure 6.1 shows the schematic of a direct current conduction pump (Nashine et al., 2007). Another type is the annular linear induction pump (ALIP), which operates on alternating current (Prashant Sharma et al., 2011). These pumps are used in the auxiliary circuits such as a purification circuit and a sodium filling circuit. Since these pumps are not an essential part of the main heat transport system, they are not discussed further.

6.3 CENTRIFUGAL PUMP

The cutaway view of a typical sodium centrifugal pump is shown in Figure 6.2. The impellor is attached to the bottom of a long shaft, while the drive and its bearings are in the upper end. To protect the bearing and avoid sodium leak from the clearances at the top, the free surface of sodium is maintained well below the bearings but sufficient to provide the suction head for the pumps. Argon, an inert gas, is used above sodium, and by the pressure control the required levels of sodium can be achieved. The centrifugal pumps are driven by variable speed drives to achieve

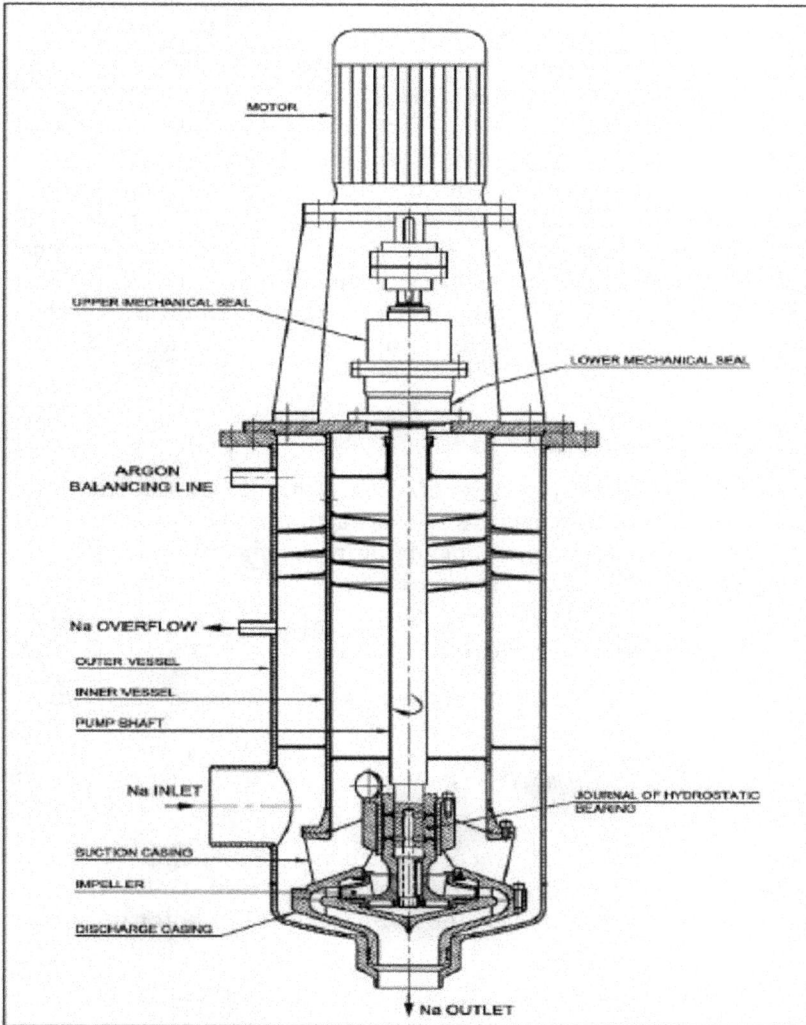

Figure 6.2 Centrifugal sodium pump.

Source: Rajan K.K, A Study on Sodium—the Fast Breeder Reactor Coolant. IOP
   Conf. Series: Materials Science and Engineering 1045, 2021.

desired sodium flows for operation at different power levels. In case of
power failure, the primary sodium pumps are provided with inertia in
the form of a flywheel on the drive shaft. This slows the flow coast-down
and provides adequate time for safety actions. Also, for short-duration
power failures, the flow reduction is not such as to need a reactor trip. As

a defense in depth, pony motors are provided along with the main drives to run the pump at minimum speed for a short duration.

The coast-down law is defined by flow-halving time, i.e., time taken for the flow to come to half the full value. This value is 8 s for PFR, PHENIX, and PFBR reactors. For RAPSODIE and FBTR it is 17 s, while for the Super Phenix reactor it is 50 s (Gouriou, 1982). Examination of the dynamics of these reactors has brought out the fact that the reactors with larger flow-halving times are designed based on no reactor trip after a power failure. This is an interesting feature. Experiments have been carried out on RAPSODIE in France (Essig et al., 1985) and EBR II in the USA (Planchon et al., 1986) for the event of power failure without reactor trip; in both cases reactor power came down because of the negative reactivity feedbacks and the plant came to a near-shutdown state.

6.3.1 Pump Hydraulic Model

The tank containing the pump essentially sees the inlet pressure to the pump at the lower end and the cover gas pressure acting on the sodium free surface at its upper end. Hence, for modeling purposes, the pump is most conveniently represented as shown in Figure 6.3.

6.3.2 Pump Dynamic Model

The torque balance equation for the shaft assembly can be written as

$$\left(\frac{2\pi}{60}\right)\left(\frac{I\Omega_D}{\Gamma_D}\right)\frac{d\alpha}{dt} = \beta_{Mt} - \beta_{Fl} - \beta_{Fr} \tag{6.1}$$

where I is the moment of inertia of coupled motor and pump rotor (kg.m^2), Ω_D is the design speed (rev/min), Γ_D is the design torque (N-m), α is the normalized pump speed, β_{Mt} is the normalized drive motor torque, β_{Fl} is

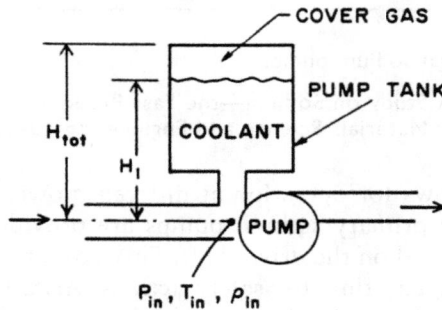

Figure 6.3 Scheme of pump hydraulic model.

the normalized hydraulic torque of the fluid, and β_{Fr} is the normalized frictional torque.

The drive motor torque goes to zero during the main motor trip or to pony motor torque during the trip to pony motor level. During normal operation the motor torque is adjusted by the action of the flow speed controllers. The fluid load torque, β_{Fl}, and the frictional torque, β_{Fr}, are determined from the pump characteristics and the flow of fluid through the pump system. The pump characteristics are defined by the head and torque versus flow as a function of speed (Stepanoff, 1957). In case of a pipe rupture, the flow through the pump may be completely in the reverse direction, as can the rotation, causing the pump to go through several regimes of operation. Therefore, analysis is required that considers the changes occurring in the characteristics as the speed and flow change. The different regimes of pump operation are illustrated in Figure 6.4. In this figure μ is the normalized flow rate and α is the normalized pump torque.

During the design stage itself it is necessary to have these characteristics. Fortunately, these characteristics are governed by the specific speed of the pump, $Ns = N\sqrt{\left(\dfrac{Q^{0.75}}{H}\right)}$, where N is the rated speed in rpm, Q the

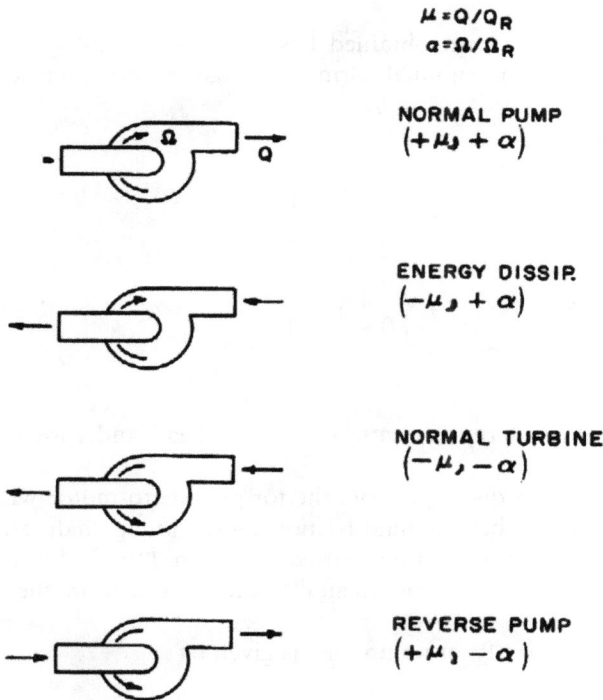

$$\mu = Q/Q_R$$
$$\alpha = \Omega/\Omega_R$$

NORMAL PUMP
$(+\mu, + \alpha)$

ENERGY DISSIP.
$(-\mu, + \alpha)$

NORMAL TURBINE
$(-\mu, -\alpha)$

REVERSE PUMP
$(+\mu, - \alpha)$

Figure 6.4 Pump configuration under different operation regimes.

discharge rate in m^3/s, and H head in m. Pumps with nearly the same specific speed have similar characteristics. If pump head, torque, flow rate, and speed are divided by their respective rated values, the dimensionless parameters are written as follows:

$$h = H/H_R$$
$$\beta = \Gamma_{hyd}/\Gamma_R$$
$$\nu = Q/Q_R$$
$$\alpha = \Omega/\Omega_R$$

Homologous theory (Streeter and Wylie, 1967) gives the normalized values of the parameters with respect to their rated values for all the regimes of operation shown in Figure 6.4, as given below:

$$\left(\frac{h}{\nu^2}\right), \left(\frac{\beta}{\nu^2}\right) Vs\left(\frac{\alpha}{\nu}\right) 0 \le \left|\frac{\alpha}{\nu}\right| \le 1$$

$$\left(\frac{h}{\alpha^2}\right), \left(\frac{\beta}{\alpha^2}\right) Vs\left(\frac{\nu}{\alpha}\right) 0 \le \left|\frac{\nu}{\alpha}\right| \le 1$$

The performance data obtained based on the homologous theory has been fitted into a polynomial form for usage in computer codes (Madni et al., 1979).

$$\left(\frac{h}{\nu^2}\right) or \left(\frac{\beta}{\nu^2}\right) = \sum_{i-o}^{n} c_i \left(\frac{\alpha}{\nu}\right)^i 0 \le \left|\frac{\alpha}{\nu}\right| \le 1 \qquad (6.2)$$

$$\left(\frac{h}{\alpha^2}\right) or \left(\frac{\beta}{\alpha^2}\right) = \sum_{i-o}^{n} c_i \left(\frac{\nu}{\alpha}\right)^i 0 \le \left|\frac{\nu}{a}\right| \le 1 \qquad (6.3)$$

The homologous representation of pump head and torque is given in Figures 6.5 and 6.6.

The frictional torque represents the torque due to motor winding, bearing, and seal losses besides fluid friction on the pump shaft. Some correlations for frictional torque are reported by Madni (1979), but it is essential to determine the same experimentally and to use it in the final safety calculations.

The inlet pressure P_{in} at pump inlet is given by

$$P_{in} = P_{gas} + pgH_l + K_{yes}\left(W_{in} - W_{out}\right)\left|\left(W_{in} - W_{out}\right)\right| \qquad (6.4)$$

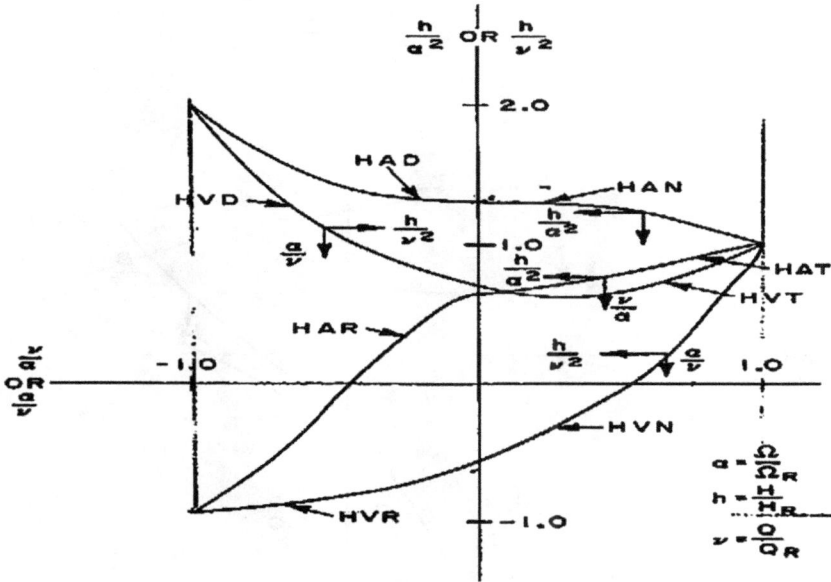

Figure 6.5 Complete homologous head curves.

Source: NUREG/CR-0240, BNL-NUREG-50859.

where Pgas is the gas pressure in the pump tank, H_l is the sodium level in the tank, W_{in}, W_{out} are, respectively, the inlet and outlet flows, and K_{yes} is a user-specified loss coefficient for the tank. Knowing P_{in} from Equation (5.4) and head developed *(PRISE)*$_{pump}$ from the homologous characteristics, the pressure at pump discharge P_{out} is simply obtained as

$$P_{out} = P_{in} + (PRISE)_{pump} \tag{6.5}$$

Mass conservation at the pump tank yields the equation describing the level of coolant-free surface:

$$A_{res} \frac{d}{dt}(\rho H_l) = W_{in} - W_{out} \tag{6.6}$$

where
  A_{res} = cross-sectional area of pump tank
  W_{in} = mass flow rate into pump
  W_{out} = mass flow rate out of pump tank
  H_l = height of coolant in pump tank
  ρ = density of coolant in pump tank

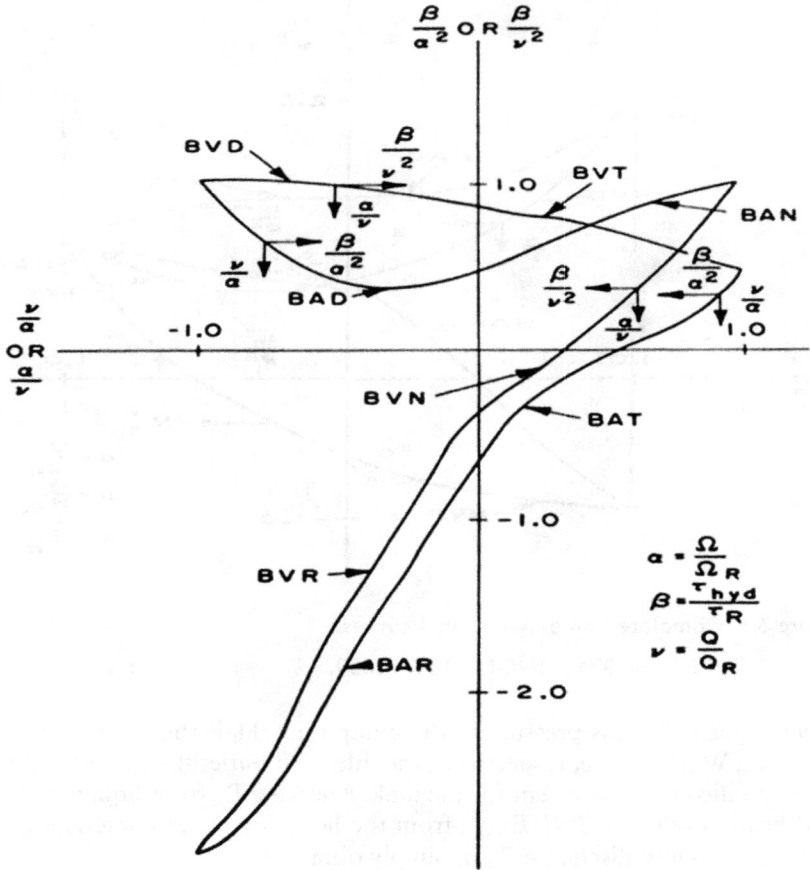

Figure 6.6 Complete homologous torque curves.

Source: NUREG/CR-0240, BNL-NUREG-50859.

6.3.3 Pump Thermal Model

For purposes of heat transfer, the pump is treated as a mixing chamber:

$$MC_M \frac{d(T)_M}{dt} = Q_i C_{Na} T_i - Q_{out} C_{Na} T_M \tag{6.7}$$

where T_i is the incoming fluid temperature and T_m is the outgoing temperature. The cover gas is treated as an ideal gas having no heat transfer interaction with sodium. Equation for cover gas pressure is given by

$$P_{gas} = (m_{gas} R_{gas} T_{gas}) / A_{res} (H_{tot} - H_l) \tag{6.8}$$

where H_{tot} is the total height of pump tank and H_l the height of coolant in the pump tank. The cover gas is assumed to be at the same temperature as sodium.

ASSIGNMENT

1. Since Electromagnetic pumps are simple and have no moving parts why do you think most of the SFRS use only Centrifugal pumps in the main sodium circuits?
2. Where is the need to have centrifugal pump characteristics for reverse flows and speeds? Under what conditions can such an event happen?

REFERENCES

Essig C., Berthet B., Gourion A., Bergonneau Ph, Vasile A. and Doneti A., (1985), "Dynamic behaviour of RAPSODIE in exceptional transient experiments", ANS Int. Topical Meeting on Reactor Safety, Knoxville, USA.

Gouriou A., (1982), "Dynamic Behavior of Super Phenix reactor under unprotected transient", Proc. LMFBR Safety Topical meeting, Lyon.

Madni I.K., Cazzoli E.G. and Agrawal A.K., (1979), "Single-phase sodium pump model for LMFBR thermal-hydraulic analysis. United States: American Nuclear Society", NUREG/CR-0240, BNL-NUREG-50859, R-7 https://www.nrc.gov/docs/ML1928/ML19289E297.pdf

Nashine B.K. Dash S.K., Gurumurthy K., Kale U., Sharma V.D., Prabhakar R., Rajan M. and Vaidyanathan G., (2007), "Performance testing of indigenously developed DC conduction pump for sodium cooled fast reactor", *Indian J. Eng. Mater. Sci.*, Vol. 14, June, pp. 209–214.

Planchon H., Sackett J.I., Golden G.H. and Sevy R.H., (1986), "Implications of EBR II Inherent Safety tests", Int. Conf. on Fast Breeders, Chalk River, USA.

Sharma, Prashant, Sivakumar L.S., Prasad R.R., Saxena D.K., Kumar V.S., Nashine B.K., Noushad I.B., Rajan K.K. and Kalyanasundaram P., (2011), "Design, development and testing of a large capacity annular linear induction pump", *Energy Procedia*, Vol. 7, pp. 622–629. https://doi.org/10.1016/j.egypro.2011.06.083

Rajan K.K., (2021), "A study on sodium - the fast breeder reactor coolant", IOP Conf. Series: Materials Science and Engineering 1045 (2021) 012013, IOP Publishing; https://doi.org/10.1088/1757-899X/1045/1/012013

Stepanoff, (1957), "*Centrifugal and Axial Flow Pumps*", John Wiley and Sons Inc., New York.

Streeter V.L. and Wylie E.B., (1967), "*Hydraulic Transients*", Mc Graw Hill, New York.

Chapter 7

Transient Hydraulics Simulation

7.1 INTRODUCTION

In the earlier chapters, various thermal models for individual components such as reactor core, pump, and IHX were described. It is essential to couple the hydraulics involving the pressure losses in the components for evolving the dynamic hydraulic model of the primary sodium system. The equations describing the hydraulic behavior of pumps, free surface levels, and flow rates in the loops together constitute the hydraulic model of a heat transport system. Since the primary and secondary sodium systems are only thermally coupled, their hydraulic models can be independently formulated and solved. In this chapter the hydraulic model of the primary and secondary sodium systems of a loop-type SFR is presented. The models are equally applicable with suitable boundary conditions for pool-type SFRs.

7.2 MOMENTUM EQUATIONS

Figure 7.1 gives the schematic of primary sodium hydraulics for a two-loop FBTR reactor.

Volume averaged momentum equations can be written relating the rate of change of mass flow to the end pressures and pressure drops. Since there are two loops in parallel, similar equations will be written for both loops. Given below are the equations for one loop. In these equations A refers to flow inertia in m^{-1}, K refers to resistance coefficient for the pipe segment, z refers to elevation, and P refers to the pressure as indicated in the Figure 7.1.

For the flow Q_{31} between reactor and IHX of loop 1,

$$A_3 \frac{dQ_{31}}{dt} = -K_{31}^2 Q_{31}^2 + \left(Z_R - Z_s\right)\rho g - \left(Z_{I1} - Z_{EE}\right)\rho g - P_1' + P_R + \Delta Z_3 \rho g \quad (7.1)$$

For the flow Q_{21} between IHX 1 outlet and pump 1,

DOI: 10.1201/9781003283188-7

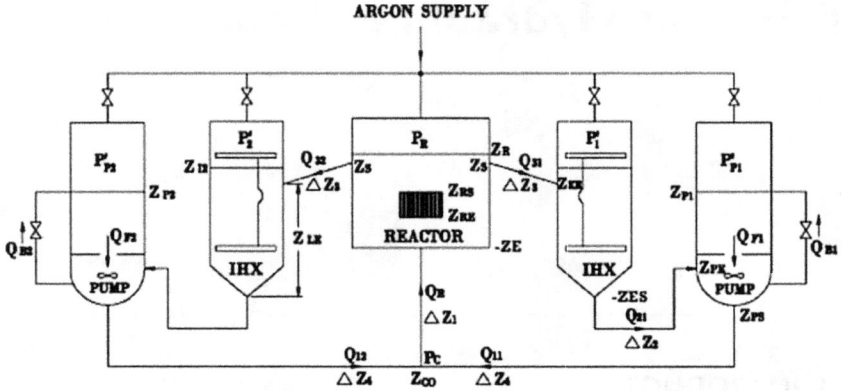

Figure 7.1 Primary hydraulic schematic of FBTR. (Vaidyanathan et.al, Dynamic Model of Fast Breeder Test Reactor. *Annals of Nuclear Energy* 37, 2010.)

Legend: P_R, $P_1{'}$, $P_{P1}{'}$ are, respectively, the cover gas pressures in reactor, IHX, pump of loop 1, and Z_R, Z_S, Z_{EE}, Z_{11}, Z_{ES}, Z_{P1}, Z_{PE}, Z_{PS}, Z_{co} are, respectively, elevations of reactor free level, reactor outlet pipe, IHX inlet, IHX free level, IHX outlet, pump free level, pump inlet, pump outlet, and reactor junction point.

$$A_2 \frac{dQ_{21}}{dt} = \left(Z_{I1} - Z_{EE}\right)\rho g + P_1{'} + Z_{LE}\rho g + \Delta Z_2 \rho g \tag{7.2}$$

$$-\left(Z_{P1} - Z_{PE}\right)\rho g - P_{P1}{'} + K_f Q_{f1}^2 - K_{21}Q_{21}^2$$

From the flow Q_{11} between pump1 to reactor inlet junction point,

$$A_1 \frac{dQ_{11}}{dt} = P_{p1}{'} + \left(Z_{p1} - Z_{ps}\right)\rho g + DPP_1 - K_f Q_{f1}^2 + \Delta Z_4 \rho g - K_{11}Q_{11}^2 - P_c \tag{7.3}$$

where DPP_1 refers to the pump developed head.

From junction point to reactor,

$$A_R \frac{dQ_R}{dt} = -K_R Q_R^2 + \Delta Z_1 \rho g - \left(Z_{RE} - Z_E\right)\rho g - DPR - \left(Z_R - Z_{RS}\right)\rho g - P_R + P_c \tag{7.4}$$

where DPR refers to the pressure drop across the reactor.

The pressure drop DPR in the reactor is common to all the zones and is given by the sum of inertial, acceleration, frictional, and potential pressure drops.

$$DPR = a_i \frac{dQ_{Ri}}{dt} + b_i Q_{ri}^2 + d_i, \left(i = 1,\ldots,5\right) \tag{7.5}$$

$$Q_R = \Sigma Q_{Ri} \qquad (7.6)$$

where a_i refers to flow inertia, b_i is the frictional drop coefficient, and d_i is the potential drop. While flow inertia is given by length divided by cross-sectional area, the other coefficients need to be initially calculated based on available correlations, but ultimately accurate data obtained from experiments must be used.

The above equation is written for all groups of parallel channels assuming that there is no radial pressure variation in the inlet and outlet of the subassemblies. This fact is based on experimental observations (Sreedhar Rao et al., 2001).

The total flow to reactor is given by

$$Q_R = Q_{11} + Q_{12} \qquad (7.7)$$

The above set of model equations would suffice for analyzing events in which the primary sodium system is intact. One of the DBEs that needs to be analyzed is the rupture of one primary pump discharge pipe. In such a case the leakage flow through the break also needs to be considered. The modeling of primary sodium system hydraulics is presented in Appendix B.

7.3 FREE LEVEL EQUATIONS

The free level in the capacities is a function of the cover gas pressure and the variation in inlet and outlet flows. Considering that the pressures are constant and maintained at the same level as the reactor cover gas pressure, the equations for IHX level can be written as

$$\frac{dZ_{I1}}{dt} = \frac{Q_{31} - Q_{21}}{\rho A} \qquad (7.8)$$

Similar equations are written for the pump level. The determination of pump levels is important to consider gas entrainment at low sodium levels. Gas entrainment can induce reactivity changes when gas-laden sodium passes through the core and can result in control problems.

7.4 CORE COOLANT FLOW DISTRIBUTION

The coolant flow distribution in the fuel, blanket, control, shielding, and other assemblies is designed for the optimum characteristic at rated operating conditions. However, in regards to the storage assembly locations where fuel assemblies are stored after they have reached a certain burnup, their heat generation (~5–30kW) is almost independent of reactor power. At rated power, these storage subassemblies will be overcooled. If the reactor is operated at conditions different from the design conditions, the fractional flow through assemblies will show deviations. This is essentially due to the differences in the buoyancy effect, which is a function of the prevailing temperature in the channels.

Fuel subassembly (SA) is composed of the foot, body, and head (Figure 7.2). The foot is a cylindrical tube with slots for coolant entry at the bottom (Prakash et al., 2011) The body is basically a hexagonal sheath housing a cluster of fuel pins in a triangular pitch.

The head of the SA is connected to the end of the body. The head houses the axial top shield in the form of a cluster of pins to prevent a neutron dose

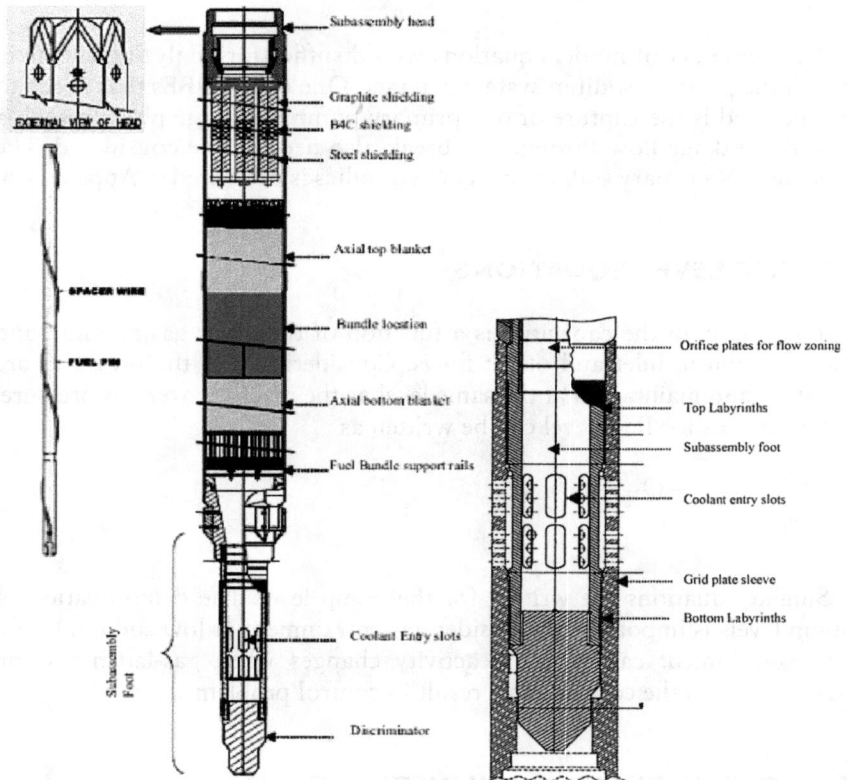

Figure 7.2 PFBR fuel SA and SA foot details. (Prakash et.al, Experimental Qualification of Subassembly Design for Prototype Fast Breeder Reactor. *Nuclear Engineering and Design* 241, 2011.)

on the components above the SA—the control plug (CP), roof slab, heat exchangers, and primary pumps—from the neutrons streaming out of the core. The shielding pin bundle also facilitates mixing of the coolant emerging out of the fuel pin bundle before leaving the SA. The coolant entering the SA through the radial slots provided in the foot (Figure 7.2) passes through the space between the fuel pins (removing the fission heat) and gets thoroughly mixed, and further flow passes the shielding bundle before leaving the SA through the head exit.

Coolant flow allocation to the SAs is based on power generation. Power generation in the core varies due to variation in the neutron flux. Also, the power generated in the SA changes from beginning-of-life to end-of-life with fuel burnup. To make the SAs power uniform, enrichment needs to be increased as we proceed from the inner to the outer rings. The blanket power increases with the burnup due to accumulation of fissile material. As the power distribution in the reactor is not uniform, the flow through each SA must be allocated such that at the outlet of the SA, the temperature is nearly uniform for all SAs. However, it is very complicated either to allocate different flows or to manage different enrichments for each SA individually. Devising many pressure drop devices to suit different SAs is a tedious process and may lead to fuel handling complications and mechanical interlocks to avoid wrong loading of SAs. So, a zone of SAs whose flow requirement values are closer can be grouped based on the maximum flow requirement of the SA in that group. Indeed, an optimized design is preferred in which some SAs having similar powers are grouped together and assigned the same flow, and few different enrichment zones are provided to get a more uniform power.

Typical arrangement of orifice plates and labyrinths at the foot of the SA with radial entry is also shown in Figure 7.2. To minimize flow leakage between the sleeve and the foot, labyrinths are provided. Stacks of orifice plates are widely used as flow-regulating devices. Orifice plates can be of machined hole design or honeycomb type. A typical honeycomb orifice plate developed for PFBR is shown in Figure 7.3. Stacking similar orifice plates with different orientations of each plate gives the different pressure drop needed. After passing through orifice plates, the flow enters a divergent foot region where the flow gets developed and then enters the fuel bundle region. The flow mixes in the bundle region because of pin spacers and finally comes out of the bundle region and then passes through the top shielding bundle and leaves axially at the head of the SA.

Correlations for the friction factor for rod bundles have been arrived at by many experimenters, namely Novendstern (1972), Rehme (1973), Chiu et.al. (1978), Engel et.al. (1979), and Cheng and Todreas (1986). An excellent review of all pressure drops correlations for a wire-wrapped fuel bundle has been presented by Bubellis and Schikorr (2008). This was further investigated based on experimental data on rod bundles in different countries

Figure 7.3 Single honeycomb orifice and stacked orifice plates. (Prakash et.al, Experimental Qualification of Subassembly Design for Prototype Fast Breeder Reactor. *Nuclear Engineering and Design* 241, 2011.)

(Chen et al., 2014). It indicates that the Cheng-Todreas correlation is the best. The detailed correlations with their applicable limits can be seen in Chen et al. (2014).

Figure 7.4 shows a comparison of the different correlations with the experimental results. There is a smooth transition from laminar to turbulent flow for wire-wrapped rod bundles, which contrasts with the more conventional friction factor for round tubes, where a discontinuity is there in the transition region. During coast-down of the pump after power failure, the flow comes down to natural convection conditions. A basic question arises about the use of steady-state friction and form loss coefficients for the full range of flow. Bishop (1980) has reviewed data on the friction factors and established that differences in form loss and friction during a transient tend to neutralize each other, as they are in opposite directions. Thus, steady-state turbulent friction factors can be used for transient conditions.

The pressure drop in the fuel and other subassembly is estimated initially using the empirical correlations, but then a full dummy subassembly is tested in water to get the accurate pressure drop. The dummy subassembly must be fabricated with the same material and other specifications as those of the reactor subassembly. This is a must so that we can get realistic value of pressure drop and can specify the required head for the pump.

7.5 IHX PRESSURE DROP CORRELATIONS

Due to the hydraulic similarity between liquid sodium and water, pressure drop correlations for water can be applied to liquid sodium. Experimental correlations to evaluate pressure loss coefficients are reported in open

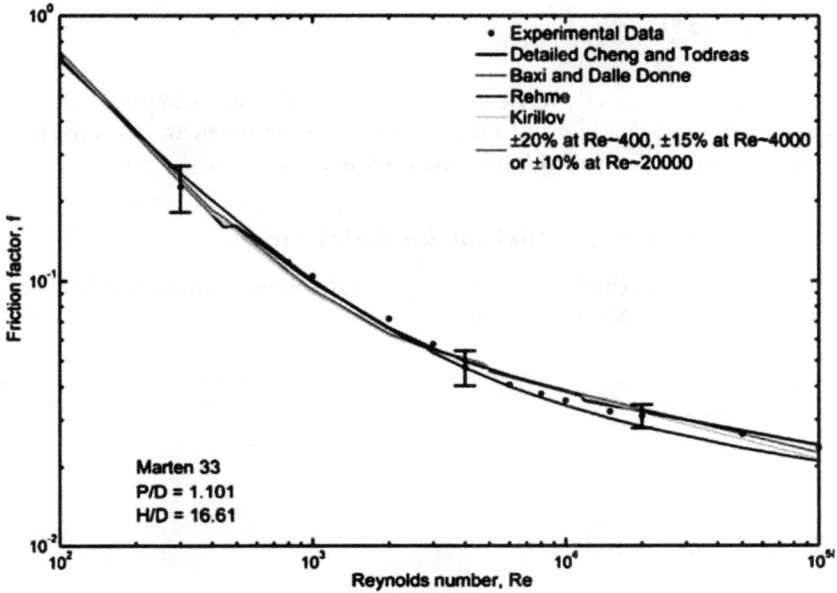

Figure 7.4 Friction factor for fuel pin bundle. (Chen et al., Evaluation of Existing Correlations for the Prediction of Pressure Drop in Wire-Wrapped Hexagonal Array Pin Bundles. *Nuclear Engineering and Design* 267, 2014.)

literature (Zukauskas and Ulinskas, 1983; Idelchek, 1966) as function of bundle geometry and Reynolds number, both for cross-flow as well as parallel flow.

7.5.1 Resistance Coefficient for Cross-Flow

This is applicable to cross-flow of primary sodium at IHX inlet. The flow resistance coefficient K for turbulent cross-flow over a bundle of smooth-wall staggered tubes is given by Idelchek (1966),

$$\Delta P = \frac{K\rho U^2}{2} \qquad (7.9)$$

where

$$K = A\,Re^{-0.27}\left(Z+1\right)$$

$$A = 3.2 + \left(4.6 - 2.7\left(\left(S1 - d\right)/\left(S2' - d\right)\right)\right)\left(2.0 - S1/d\right)$$

$$S2' = \left(0.25\,S1^2 + S2^2\right)^{0.5}$$

S_1, S_2 are, respectively, vertical and horizontal distances between the axes of adjacent tubes in a bundle, d is the tube outer diameter, and Z is the number of transverse rows of tubes in the bundle.

7.5.2 Resistance Coefficient for Axial Flow

For the axial flow, the friction factor, f, is calculated from Idelchek (1966), and pressure drop ΔP is given by

$$\Delta P = \frac{\rho f L V^2}{2 D_{eq}} \qquad (7.10)$$

where

$$f = 0.11\left(\left(\varepsilon/d\right) + \left(68/Re\right)\right)^{0.25}$$

In the above, ρ refers to density, L to the length of the tube, V to the mean flow velocity, D_{eq} to the equivalent tube diameter, d to tube diameter, ϵ to tube roughness, and Re is the Reynold's number.

7.6 PUMP CHARACTERISTICS

The solution of the transient hydraulics of the sodium system involves the pump head–flow-torque characteristics. These are presented in detail in Chapter 6.

7.7 COMPUTATIONAL MODEL

The set of differential equations of the primary sodium system comprise the flow, pump speed, and level equations. While the time constants of the level equations and pump speed are high due to large volume capacities, the flow equations have a low time constant due to low flow inertias. Thus, the set of equations become a set of stiff differential equations. Application of finite differencing schemes for solution poses problems due to high nonlinearity of the flow and pressure drop unlike the thermal equations. For solution of such equations available solvers are the Runge Kutta Gill (Ralston Wilf, 1960), Runge Kutta Merson (William, 1973), Gear's method (Ralston Wilf, 1960), and Hamming's Predictor Corrector Method (Ralston Wilf, 1960). It has been reported that Runge Kutta Gill fails for a high degree of stiffness

(Hopper, 1973). The Hamming's Predictor Corrector Method has been seen to be computationally economical compared to Gear's and Runge Kutta Merson Methods (Vaidyanathan and Kothandaraman, 1979). For speeding up the numerical calculations, the above equations are split Into two sets. One set comprises the equation of IHX flow, pump flow, levels, and pump speeds, and the other set comprises the equations of core channel flows. The first set is solved through a standard ODE solver based on the Hamming's Predictor Corrector method. In the second set involving core channel flows, the inertial drop term in the momentum equations can be neglected, as it is small compared to the other drops. With this simplification one gets a set of algebraic equations. For the solution one needs the pressure drop across the core for which the previous time step value is used. In this manner faster solution can be achieved.

7.8 VALIDATION STUDIES

Modeling of the hydraulics is very crucial to the determination of temperatures. For accurate evolution of flow, proper modeling of inertia of rotating systems and inertia of fluid is essential. This aspect is verified for the event of primary sodium pumps trip. Predictions made by DYNAM for FBTR are compared with experimental results measured at the plant (Vaidyanathan et al., 2010). Figure 7.5 shows the coast-down of the east and west loop

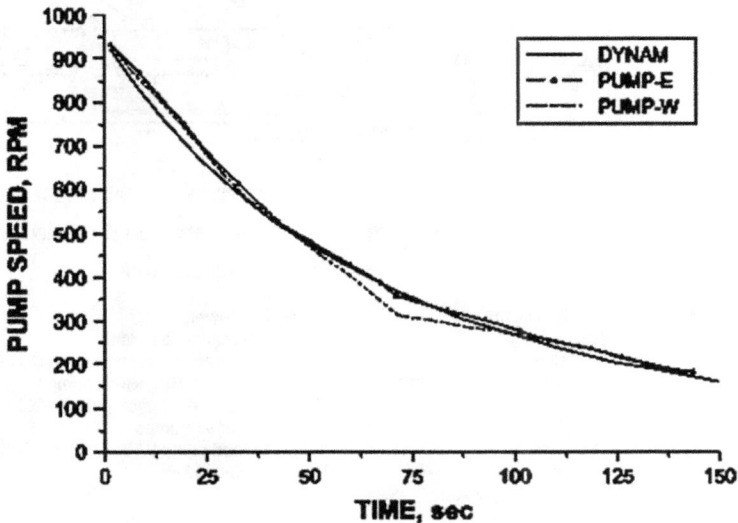

Figure 7.5 Primary pump coast down in FBTR. (Vaidyanathan et.al., Dynamic Model of Fast Breeder Test Reactor. *Annals of Nuclear Energy* 37, 2010.)

primary pumps as a function of time. There is a very good match between the measured and computed speeds validating the primary hydraulics model.

7.9 SECONDARY CIRCUIT HYDRAULICS

Figure 7.6 shows the schematic of the FBTR, a loop-type reactor in India (Srinivasan et al., 2006). The secondary sodium exiting the IHX passes

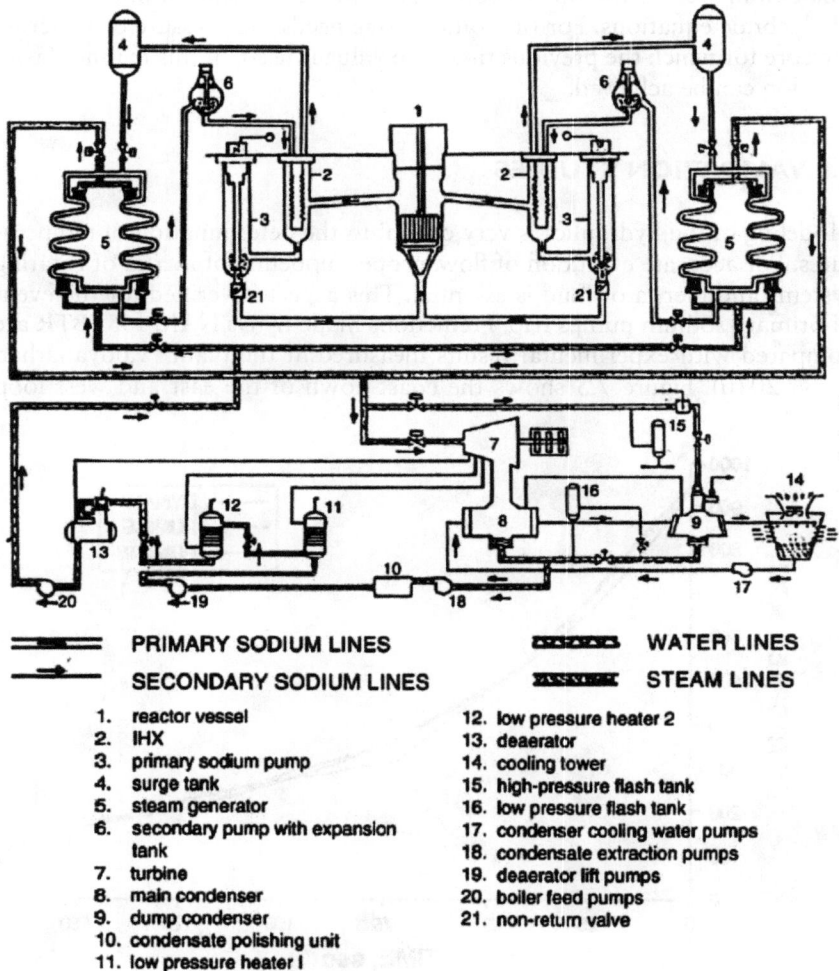

PRIMARY SODIUM LINES

SECONDARY SODIUM LINES

WATER LINES

STEAM LINES

1. reactor vessel
2. IHX
3. primary sodium pump
4. surge tank
5. steam generator
6. secondary pump with expansion tank
7. turbine
8. main condenser
9. dump condenser
10. condensate polishing unit
11. low pressure heater I

12. low pressure heater 2
13. deaerator
14. cooling tower
15. high-pressure flash tank
16. low pressure flash tank
17. condenser cooling water pumps
18. condensate extraction pumps
19. deaerator lift pumps
20. boiler feed pumps
21. non-return valve

Figure 7.6 FBTR flowsheet. (Srinivasan et.al, The Fast Breeder Test Reactor— Design and Operating Experiences. *Nuclear Engineering and Design* 236, 2006.)

through pipes to a surge tank and from there is led to the steam generator through pipes. Sodium coming out of the steam generator is sent through pipes to the pump tank. Sodium is pumped from the pump tank to the IHX through piping. In case of a leak in the steam generator tubes there is a likelihood of high-pressure water entering sodium and causing an exothermic reaction resulting in high pressures due to hydrogen generation. The purpose of providing two sodium capacities with free levels is to absorb the pressure surge generated in the steam generator due to a large sodium–water reaction. The expansion tank also helps accommodate the volumetric expansion of sodium with rise in temperatures from shutdown state to full power. Under forced flow conditions the sodium flow in all the sections of secondary sodium circuit will be same. However, under natural convection conditions, the level changes in the two tanks will also influence the flows. Hence, the secondary sodium circuit is modeled in two segments: one section from the pump tank to the surge tank through IHX, and the other section from the surge tank to the pump tank through the steam generator to the pump tank.

7.9.1 Secondary Hydraulics Model

For the segment 1 from the pump tank to the surge tank, the momentum balance yields Equation 7.12, where $\int \rho g z$ is the buoyancy term.

$$I_1 \frac{dW_1}{dt} = P_{st} - P_{pt} - \Delta P_{f1} - \int \rho g z_1 \tag{7.11}$$

where W is the flow, I is the flow inertia (length/cross section area of pipe), P is cover gas pressure, ΔP is pressure drop, and z is the elevation difference of a pipe segment, while subscripts $_1$, st, pt, f, and g refer, respectively, to the pipe segment, surge tank, expansion tank, frictional pressure drop, and gravitational pressure drop in the pipe segment.

For the segment 2 from the surge tank to the pump tank, the momentum balance yields.

$$I \frac{dW_2}{dt} = P_{pt} - P_{st} + \Delta P_{pt} - \Delta P_{f2} - \int \rho g z_2 \tag{7.12}$$

Level equation for surge tank is given by

$$(\rho A)_{st} \frac{dL_{st}}{dt} = W_2 - W_1 \tag{7.13}$$

Level equation for pump tank is given by

$$(\rho A)_{pt} \frac{dL_{pt}}{dt} = W_1 - W_2 \tag{7.14}$$

where Lst and Lpt are the levels in surge tank and expansion tanks, ρ is density of sodium, and A is the cross-section area of sodium tanks.

The above model is referred to as a two-segment model in further discussions. If we do not consider the level changes, one can represent the whole secondary circuit by a single equation, as flows throughout will be same. This is referred to as a single-segment model. The equations are solved by the Hamming's Predictor Corrector method as in the case of primary hydraulics equations.

7.9.2 Natural Convection Flow in Sodium Validation Studies

Tests were done in the FBTR reactor at low power of 180 kW. The heat sink for this event are the piping heat losses and natural air cooling of the surface of SG. With steady operation at 180 kW, the two secondary pumps were tripped but the reactor was not tripped. The secondary flow comes down to minimum within a few minutes and then rises as governed by the buoyancy forces. The development of natural convection flow as measured in the plant is depicted in Figure 7.7. To simulate this event, the SG casing heat transfer model developed above was integrated with the secondary and primary circuit models. The flow of secondary sodium oscillates and stagnates from 15 to 75 min in the test.

Predictions by the two-segment model (Figure 7.7) are not able to reproduce the initial oscillations, observed in the test, but then the final steady

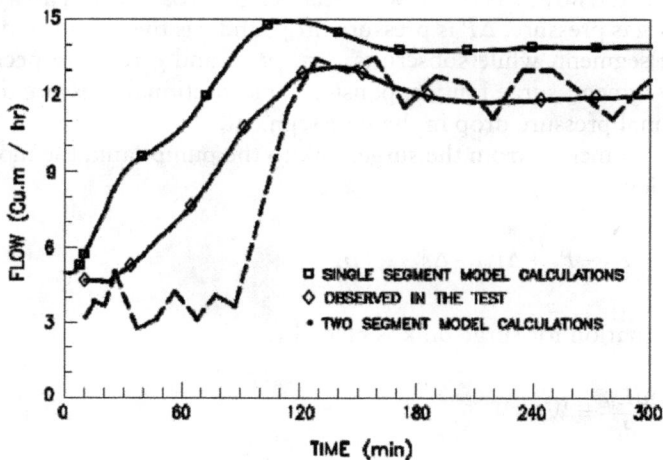

Figure 7.7 Natural circulation flow in secondary sodium – FBTR. (Vaidyanathan G. et al., Natural Convection in Secondary Sodium Circuit of Fast Breeder Test Reactor. *International Journal on Design and Manufacturing Technologies* 5(1), 2011.)

state flow is matching with test data. This can be explained by the thermal stratification in the piping, which is not modeled effectively in the one-dimensional model, or the effect of pressure drop coefficients used. To assess the effect of neglecting the surge and pump tank level modeling and treating the secondary loop as a single segment, the calculations were carried out (Figure 7.7). It is seen that the flows by a two-segment model are closer to reactor data and thus need to be used.

ASSIGNMENT

1. For typical centrifugal pump characteristics, draw the characteristics with one pump and two pumps in parallel.
2. What are the difficulties in solving all equations of primary sodium hydraulics involving levels and flow together? Attempt a computer program with two flow equations and two capacity-level equations and try to solve them using differential equation solvers, and present the issues involved.

REFERENCES

Bishop A.A. (1980), "Transient friction and form loss factors in turbulent and laminar flow: a review", *Nuc. Eng. Des.* Vol. 62, p. 361.

Bubellis E. and Schikorr M. (2008), "Review and proposal for best fit of wire wrapped fuel bundle friction factor and pressure drop predictions using various existing correlations", *Nuc. Eng. Des.*, Vol. 238, p. 3299.

Chen, S.K., Todreas, N.E. and Nguyen, N.T. (2014), "Evaluation of existing correlations for the prediction of pressure drop in wire-wrapped hexagonal array pin bundles", *Nuc. Eng. Des.*, 267, 109–131. https://doi.org/10.1016/j.nucengdes.2013.12.003.

Cheng S.K. and Todreas N.E. (1986), "Hydrodynamic models and correlations for bare and wire wrapped hexagonal rod bundles- bundle friction factors, sub channel friction factors and mixing parameters", *Nuc. Eng. Des.*, Vol. 92, p. 227.

Chiu, C., Rohsenow, W.M. and Todreas, N.E., (1978), "Turbulent sweeping flow mixing model for wire wrapped LMBFR assemblies, COO-2245-55TR," Rev. 1, MIT.

Engel F.C., Markley R.A. and Bishop A.A. (1979), "Laminar, transition and turbulent parallel flow pressure drop across wire wrap spaced rod bundles", *Nucl. Sci. Eng.*, Vol. 69, p. 290.

Gajapathy R., Velusamy K., Selvaraj P., Chellapandi P. and Chetal S.C., (2009), "A comparative CFD investigation of helical wire-wrapped 7, 19 and 37 fuel pin bundles and its extendibility to 217 pin bundle", *Nucl. Eng. Des.*, 239, pp. 2279–2292.

Hopper M.J. (1973), "Harewell subroutine library-a catalogue of subroutines", AERE-R-7477.

Idelchek, I.E. (1966), *Handbook of Hydraulic Resistances*, AEC-TR-6630.

Novendstern E.H. (1972), "Turbulent flow pressure drop model for fuel rod assemblies utilizing helical wire wrap space system", *Nuc. Eng. Des.*, Vol. 22, p. 19.

Prakash V., Thirumalai M., Anandaraj P., AnupKumar, Ramdasu D., Pandey G.K., Padmakumar G., Anandbabu C., Kalyanasundaram P. (2011). Experimental qualification of Subassembly Design for Prototype Fast Breeder Reactor, *Nuclear Engineering and Design* 241.

Ralston, Anthony and Wilf S. (1960), "*Mathematical Methods for Digital Computers*", John Wiley, New York, Vol. 1.

Rehme K. (1973), "Pressure drop correlations for Fuel element spacers", *Nucl. Technol.*, Vol. 17, p. 15.

Sreedhar Rao M.V., Padma Kumar G., Vaidyanathan G., Prabhakar R., Ghosh D., Govindarajan S. and Kale R.D. (2001), Flow distribution in the grid plate of prototype fast breeder reactor at different operating conditions, National Conference on Fluid Machines and Fluid Power, Chandigarh, India.

Srinivasan G., Kumar, K.S., Rajendran, B. and Ramalingam, P.V. (2006), "The fast breeder test reactor—design and operating experiences", *Nucl. Eng. Des.*, 236.

Vaidyanathan G., Kasinathan N. and Velusamy K. (2010), "Dynamic model of fast breeder test reactor", *Ann. Nucl. Energy*, 37, 450–462. http://dx.doi.org/10.1016/j.anucene.2010.01.013

Vaidyanathan G., Kasinathan N. and Velusamy K. (2011), "Natural convection in secondary sodium circuit of fast breeder test reactor", *Int. J. Des. Manuf. Technol.*, Vol. 5, No. 1, January, pp. 1–7.

Vaidyanathan G. and Kothandaraman A.L. (1979), "Flow transients in FBTR", *Proc. Power Plant Safety and Reliability*, Department of Atomic Energy, Mumbai, India.

William P.W. (1973), "*Numerical Computations*", Thomas Nelson, London.

Zukauskas A. and Ulinskas R. (1983), "*Heat Exchanger Design Handbook*", Hemisphere Publishing Corporation, Washington.

Chapter 8

Steam Generator

8.1 INTRODUCTION

A steam generator is an important component of a fast reactor plant. Hot sodium coming out of IHX enters the steam generator through a nozzle at the top and flows down on the shell side. Water at high pressure and low temperature enters at the bottom end and comes out as high-pressure, high-temperature steam at the top. The presence of sodium and water separated by a tube wall has the potential for a sodium–water reaction in case of a tube leak. Experience so far has indicated that most of the leaks occur at the tube-to-tube sheet welds, and the consequences can be handled in a safe manner by suitable design. The reaction between water and sodium produces sodium hydroxide and hydrogen at high temperature, as it is an exothermic reaction. The generated hydrogen at the reaction site pushes the sodium away from the reaction site, resulting in a pressure surge in the secondary sodium system. To take care of this, two tanks with enough sodium capacity and free levels with argon cover gas are provided both upstream and downstream of the steam generator. This chapter is devoted to a brief description of the steam generators used and the dynamic model of the steam generator.

8.2 HEAT TRANSFER MECHANISMS

Consider a vertical tube heated uniformly along its length as shown in Figure 8.1 (Vaidyanathan, 2013). While the liquid is being heated up to its saturation temperature at the local pressure at that height in the tube, the wall temperature initially is below that necessary for nucleation. Thus, the heat transfer process is subcooled, single-phase heat transfer to the liquid, which may be laminar or turbulent. Then, the wall temperature rises above the saturation temperature, and boiling nucleation takes place in the superheated thermal boundary layer on the tube wall, such that subcooled flow boiling occurs with the vapor bubbles condensing as they drift into the subcooled core (Collier and Thome, 1994). The liquid then reaches its

DOI: 10.1201/9781003283188-8

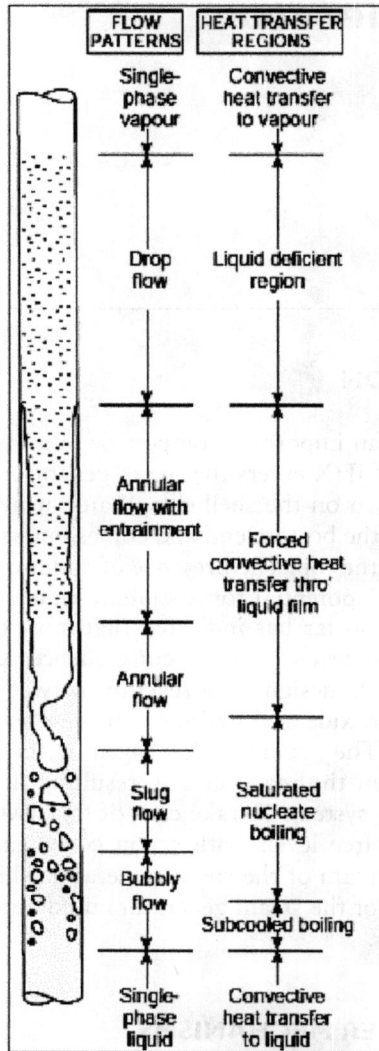

Figure 8.1 Heat transfer regimes. (Vaidyanathan, G., *Nuclear Reactor Engineering—Principles and Concepts*. Delhi, India: S. Chand Publishing, 2013.)

saturation temperature, and saturated boiling in the form of bubbly flow begins. Saturated boiling continues through the slug flow regime, the annular flow regime, and then the annular flow regime with liquid entrainment in the vapor core. Subsequently, the annular film is either dried out or sheared from the wall by the vapor, a point that is referred to as the onset of dry-out or simply dry-out. Above this point, referred to as post dry-out, the mist flow in the form of entrained droplets occurs with a large increase in wall temperature. The temperature of the continuous vapor phase in this region tends to rise above the saturation temperature, and heat transfer occurs via four mechanisms: single-phase convection to the vapor, heat transfer to the droplets within the vapor, heat transfer to droplets impinging on the wall, and thermal radiation from the wall to the droplets. When all the liquid has evaporated, the heat transfer is by single-phase convection to the dry vapor.

8.3 STEAM GENERATOR DESIGNS

It is worth to go through some basics of conventional boilers/steam generators before getting into the details of steam generators used in sodium-cooled fast reactors.

8.3.1 Conventional Fossil-Fueled Boilers

The feedwater used in power stations contain large amounts of dissolved silicates. It volatilizes (becomes a gas) at the high pressures and temperatures within the boiler. Steam cools as it moves through the turbine. At these lower temperatures, silica precipitates onto the turbine blades where it accumulates as a glassy deposit. As the silica deposits accumulate on the turbine blades, they cause a pressure drop within the turbine itself. The deposits are of random thickness and cause balance and vibration problems inside the turbine. Silica deposits on the blades and other elements of turbine restrict steam flow to it from the boiler. This results in a loss of output from the turbine and a reduction in the turbine's electricity generation capacity.

8.3.1.1 Drum Type

Drum-type boilers have been used generally in conventional coal-fired power plants (Figure 8.2). Feedwater picks up the residual heat of the flue gases in the economizer and enters the drum. From the drum the feedwater goes down to enter the boiler from the bottom. The design is such that water reaches saturation temperature and reaches a steam quality of ~ 0.3 (below dry-out as in Figure 8.1). The idea is to avoid region G (Figure 8.1), in which heat transfer coefficients of water–steam mixture are low, leading to high tube temperatures. Then the steam–water mixture goes to the drum where the steam is separated from the water. The separated steam then goes to the superheater. Due to its large water content, when there is sudden demand for more steam, the water can flash into steam and thus provide a load-following capability to the power plants. The drum type is also suitable from the considerations of water quality maintenance by allowing impurity separation in the drum and having blowdown at different times to remove

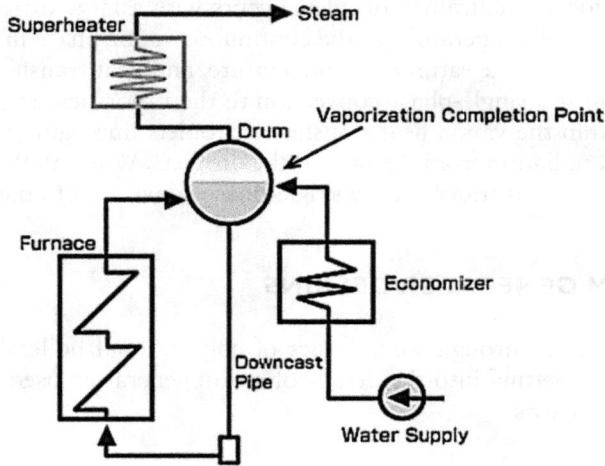

Figure 8.2 Schematic drum-type boiler.

the impurities that separate out in the drum. The major aspect is to have silica <1 ppm in the steam going from the drum to the superheater. The blowdown provision is fine if the pressure of the boilers is low. The allowable silica comes down as pressure increases. With pressures greater than ~10 Mpa, it becomes necessary to demineralize the boiler water after the condenser. These units consist of cation and anion polishing units that selectively remove the various impurities.

8.3.1.2 Once Through Steam Generators (OTSG)

As the thermodynamic cycle efficiency is high with higher pressures and temperatures, the boiler drum needed higher thicknesses, and interest arose to do away with the drum. The first significant commercial application of once-through boilers (Figure 8.3) was made by Mark Benson, a Czechoslovakian inventor, when in 1923 he provided a 4 ton/h unit for English Electric Co., Ltd. at Rugby, England. In the 1930s and 1940s, power plant operating conditions were limited to the subcritical regime because of limitations of metallurgy and water chemistry control technology.

In Europe, boiler technology followed the once-through philosophy. This was driven by material availability constraints and took advantage of the fact that the once-through boiler generally used thinner-walled tubes than that needed for a drum. Thus, the once-through boiler eliminated the need for thick steel plate for the steam drum. Studies have shown that while drum-type SG can take a larger step change in load due to its huge thermal capacity, the OTSG can take only a smaller step change. However, to come back to equilibrium before taking up the next load change, the drum type

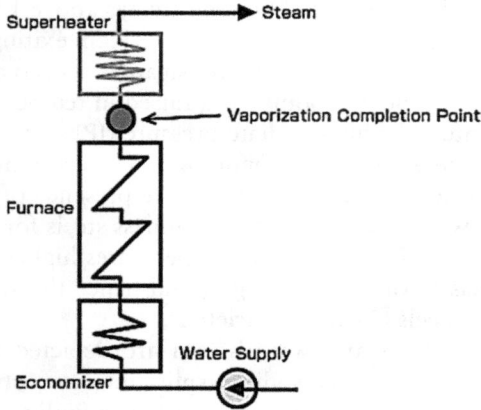

Figure 8.3 Once-through steam generator schematic.

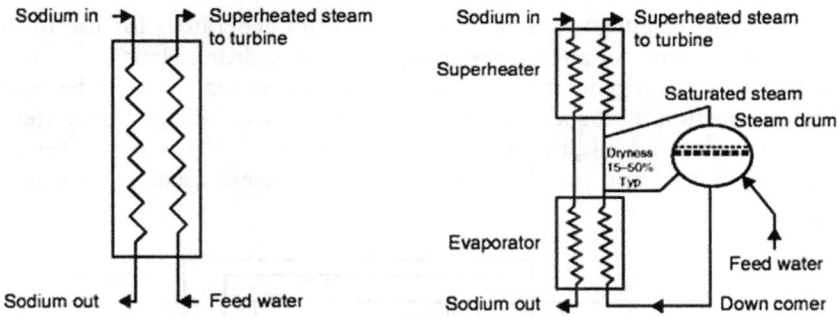

Figure 8.4 SFR steam generator concepts—OTSG and drum type.

will take more time in view of large thermal capacity involved, whereas the OTSG is ready for the next change in a shorter time.

8.3.2 Sodium-Heated Steam Generators

Initial SFRs, such as DFR and PFR in the UK, EBR-II in the USA, and BN-350 in the former Soviet Union, used the drum-type boilers that were used in the conventional power plants, for which enough operational experience existed at that time. Figure 8.4 shows the schematic of both drum-type and OTSG configurations in SFR. Secondary sodium coming out of IHX gives enough heat to the water in the steam generator to convert it to steam. In the evaporator portion, the water is heated to slightly above saturation temperature. It then enters the steam drum. Steam separated in the drum is led to the superheater, which is again heated by sodium. Superheated steam (160 b, 538°C) enters the high-pressure (HP) turbine, where it expands to nearly 40 bars. Quality of steam should be always above 12% from erosion considerations of water droplets in steam. The steam exiting the HP turbine approaches the 12% limit. Hence this wet steam is sent to a reheater, which is again heated by secondary sodium and raised in temperature to ~530°C. The steam then enters the intermediate-pressure (IP) and low-pressure (LP) stages of the turbine as shown in Figure 8.5. The economizer-evaporators were made of ferritic steel (2.25Cr-1Mo). The presence of chlorine in feed water did not allow the use of austenitic stainless steels for evaporators due to stress corrosion cracking issues. At temperatures higher than 500°C, the 2.25 Cr-1Mo steels have lower strength, and hence the superheaters were made in austenitic steels (316, 321 varieties).

Flow paths of sodium and water/steam are depicted schematically in Figure 8.5. The hot secondary sodium splits into two streams, with one entering the superheater and the other the reheater. Sodium exiting from the superheater and the reheater mix and then enters the evaporator. Steam–water mixture coming out of the HP turbine is sent to the reheater where it is superheated and sent to IP turbines.

One of the disadvantages of drum-type steam generators for use in the SFR is the presence of a large water capacity in the drum. Hence, in case of a leak in the steam generator tube, a large amount of water would be available for reaction with sodium, resulting in long-term damage to the steam generator. To avoid this, the later plants adopted the OTSG designs. Ferritic steel evaporators and stainless steel superheaters were used. The steam at

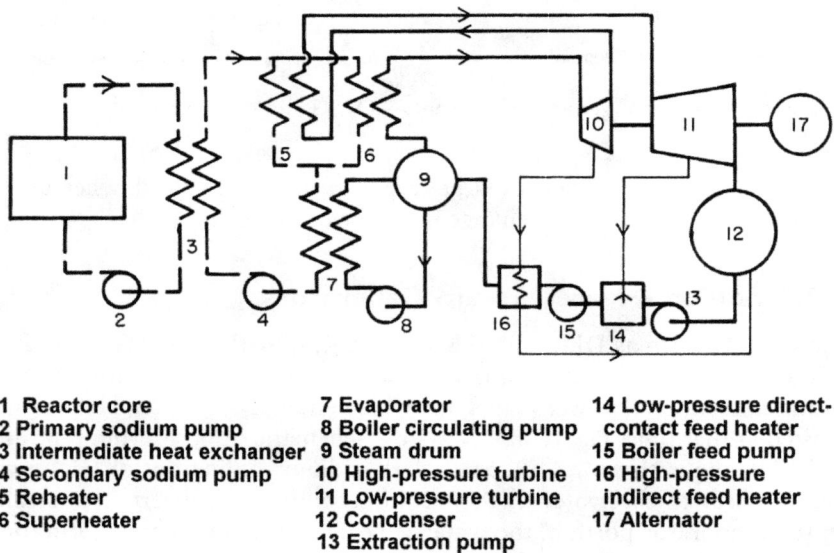

1 Reactor core	7 Evaporator	14 Low-pressure direct-
2 Primary sodium pump	8 Boiler circulating pump	contact feed heater
3 Intermediate heat exchanger	9 Steam drum	15 Boiler feed pump
4 Secondary sodium pump	10 High-pressure turbine	16 High-pressure
5 Reheater	11 Low-pressure turbine	indirect feed heater
6 Superheater	12 Condenser	17 Alternator
	13 Extraction pump	

Figure 8.5 Sodium reheat cycle schematic. (Judd, AM. *An Introduction to the Engineering of Fast Nuclear Reactors*. Cambridge University Press, 2014).

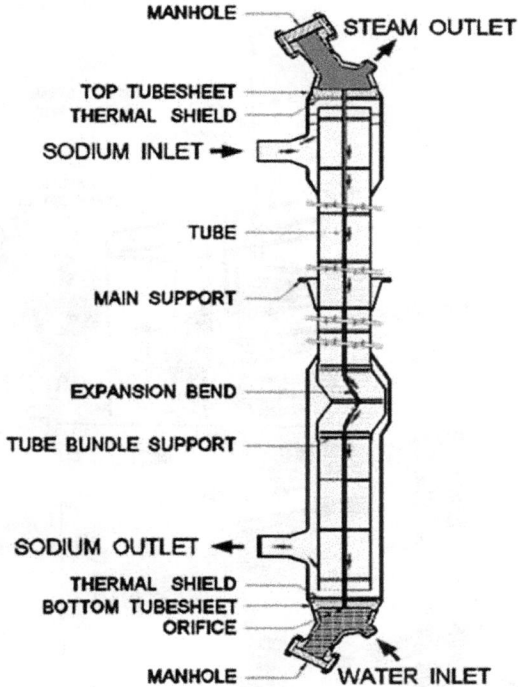

Figure 8.6 Integrated steam generator schematic.

the outlet of evaporator was slightly superheated (~20°C) to ensure that no water particles were carried out into the superheaters.

With the development of high-Cr steels like 9Cr-1Mo, both the super-heater and the evaporator in a once-through type could be integrated into a single unit (Figure 8.6).

PHENIX in France utilized the once-through design (Figure 8.7), and PFR in the UK used a drum-type design (Figure 8.5), though both were built around the same time. These two were prototype reactors that provided extensive feedback experience for future designs. Both used the sodium reheat design, as the available turbine units at that time had the same main steam and reheat steam temperatures. Since sodium was used for reheating, the steam from the outlet of an HP turbine had to be brought to the steam generator building. Taking the secondary sodium outside the steam genera-tor building was out of question, as then it would require facilities to take care of sodium leaks fire and other such incidents in the turbine building as well. This gave rise to the concept of live steam reheat, wherein the reheater could be placed in the turbine building and live steam from the steam gen-erator itself could be used for reheating the steam. This also presented the incentive to combine the evaporator and the superheater into a single

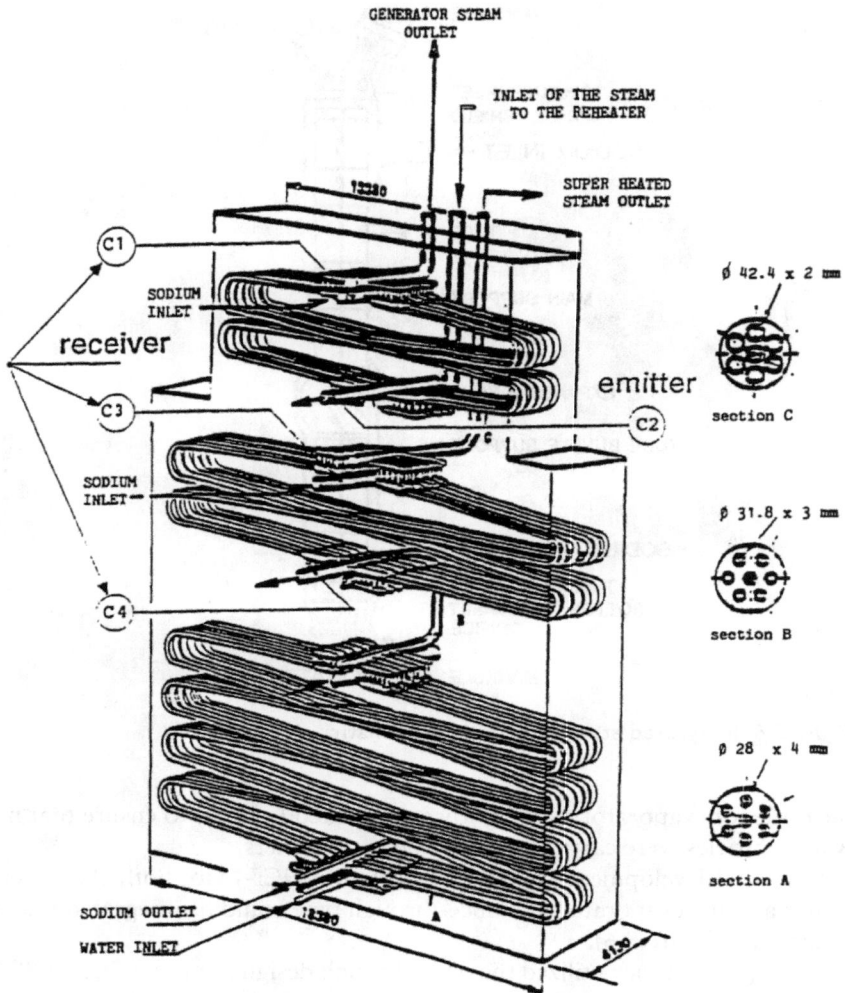

Figure 8.7 PHENIX steam generator. (International Atomic Energy Agency, International Working Group on Fast Reactors, Proceedings of the Specialists' Meeting on Acoustic/Ultrasonic Detection of In Sodium Water Leaks on Steam Generators, IWGFR/79, Aix-En-Provence, France, October 1990.)

integrated unit. This was made possible with the development of high Cr-Mo steels (9Cr-1Mo, 12Cr-1 Mo, Incolloy-800), which could be used for both regions.

With developments in turbine technology with modular designs, turbines are available to suit the live steam reheat cycles. Though there would be a reduction in cycle efficiency with lower reheat temperatures, it would be

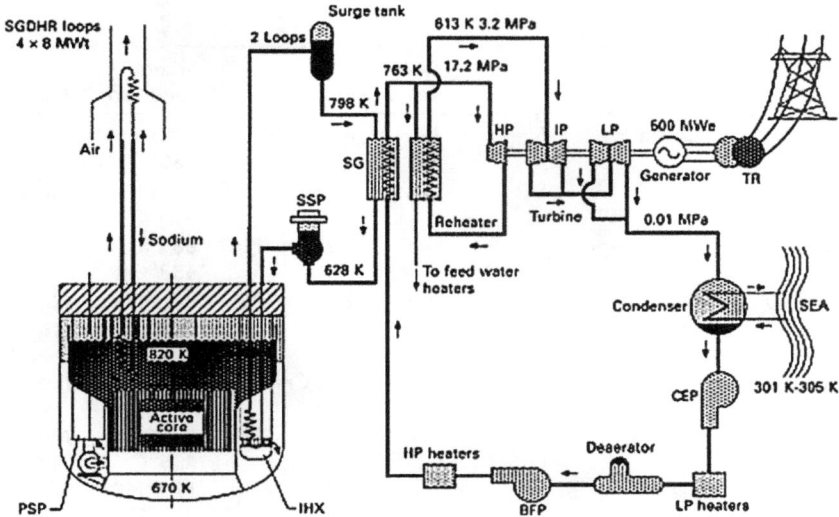

Figure 8.8 Steam reheat cycle in PFBR.

offset to a large extent by the capital cost involved in running a steam line from the turbine building to the steam generator building and back.

Experiences with all sodium-heated steam generators in PFR and PHENIX (Chetal and Vaidyanathan, 1997) have indicated that the water/steam leak into sodium occurs mainly in the weld areas at the tube-to-tube sheet junction. Hence, one of the criteria recommended for sodium-heated steam generators is to have minimum weld joints, which is achievable with very long seamless tubes.

Super PHENIX 1, a 1200-MWe reactor built in France, utilized the live steam reheat cycle. The steam generator was an integrated one with 90-m-long helical coils. PFBR, in India, is pursuing a similar approach. Figure 8.8 shows the steam reheat cycle incorporated in PFBR, India. Table 8.1 gives the list of steam generator designs used by different countries. Unlike IHX, there is diversity in steam generator concepts.

8.4 THERMODYNAMIC MODELS

An important part in the thermohydraulic analysis of a fast breeder reactor involves the determination of thermodynamic states of the steam/water cycle during steady-state and transient modes of operation. The type of thermodynamic model used to describe a steam generator system is dictated by the system geometry and the problem under consideration. Due to the complex nature of the two-phase and compressible flows in the steam generating system, a purely analytical, general solution is not possible. Hence, one must

Table 8.1 Steam generator concepts in SFRs

Plant	Concept	Tube shape	Cover gas	Material	Steam condition	Reheat
DFR	Drum type	S	No	Austenitic	1 MPa 280°C	No
FERM	OTSG	S	Yes	2.25Cr-1Mo	6.3 MPa 410°C	No
EBR-II	Drum type	DW St	No	2.25Cr-1Mo	88 MPa 438°C	No
BOR-60	Drum type, OTSG, Inverse	S	No	2.25Cr-1Mo	8.8 MPa 440°C	No
KNK-II	OTSG (Integral)	S	No	2.25Cr-1Mo Nb	6.3 MPa 485°C	No
BN-350	Drum type	B-EVA U-SH	Yes	2.25Cr-1Mo	5 MPa 435°C	No
Phenix	OTSG	S	No	2.25Cr-1Mo (EVA) 321 (SH, RH)	17.4 MPa 512°C	Sodium
PFR	Drum type	U	Yes	2.25Cr-1Mo +Nb+ Ni (EVA) 316 SH, RH	17.1 MPa 512°C	Sodium
BN-600	OTSG	St	No	2.25Cr-1Mo EVA, SS304 SH	13.7 MPa 505°C	Sodium
SNR-300 (not operated)	OTSG	H, St	No	2.25Cr-1Mo +Nb+ Ni	16.7 MPa 500°C	Steam
CRBRP (not operated)	Drum type	Hockey Stick	No	2.25Cr-1Mo (EVA+SH)	10 MPa 482°C	No
SPX-1	OTSG (Integral)	H	Yes	Alloy 800	18.2 MPa 490°C	Steam
FBTR	OTSG (Integral)	S	No	2.25Cr-1Mo +Nb+ Ni	12.5 MPa 480°C	No
MONJU	OTSG	H	Yes	2.25Cr-1Mo (EVA), Austenitic SS (SH)	12.7 MPa 487°C	No
DFBR (proposed)	OTSG (Integral)	H	No	Modified 9Cr-1Mo	17.2 MPa 497°C	Steam
EFR (proposed)	OTSG (Integral)	St	No	Modified 9Cr-1Mo	18.5 MPa 490°C	Steam
PFBR	OTSG (Integral)	St	No	Modified 9Cr-1Mo	17.7 MPa 493°C	Steam

B, bayonet; DW, double wall (EBR II double wall concept. All others single wall.); EVA, evaporator; H, helical; OTSG, once-through steam generator; RH, reheater; S, serpentine; SH, superheater; St, straight; U, U-tube.

resort to numerical solutions of the governing equations. Analysis of any flow system involves the solution of mass, momentum, and energy equations. There are four unknowns in a single-phase flow, which are velocity, pressure, temperature, and density. The number of unknowns in a two-phase flow is eight. These are: void fraction, liquid and vapor phase velocities, liquid and vapor phase densities, pressure, and the temperatures of each phase. It is apparent that the two-phase compressible flow analysis is more complicated than its single-phase counterpart. Most current simulation models are descendants of the original RELAP (RELAP3B 1974) and/or FLASH (Porshing et al., 1969) computer programs. Both models solve the conservation equations of mass, energy, and momentum for a staggered node-flow path representation of a thermal hydraulic system. These conservation laws for a one-dimensional (no slip) flow is written below (Meyer, 1961).

Mass Continuity: $\quad \dfrac{\partial}{\partial t}(\rho A) + \dfrac{\partial}{\partial x}(\rho u A) = 0$ (8.1)

Momentum: $\quad \dfrac{\partial u}{\partial t} + u\dfrac{\partial u}{\partial x} + \dfrac{1}{\rho}\dfrac{\partial p}{\partial x} + g\dfrac{\partial z}{\partial x} + \dfrac{f}{di}\dfrac{u|u|}{2} = 0$ (8.2)

Energy: $\quad \rho\left(\dfrac{\partial e}{\partial t} + u\dfrac{\partial e}{\partial x}\right) = \dfrac{\partial q}{\partial y} - p\dfrac{\partial u}{\partial x} + \mu\varphi + q'''$ (8.3)

where ρ is the fluid density, u is the velocity, A is the flow cross-sectional area, P is the pressure, g is the gravitational acceleration, f is the friction factor, z is the vertical elevation, di is the inner diameter of the channel, e is the specific internal energy, q''' is the heat generation rate, q is the heat transfer rate, and Φ is the dissipation function.

Defining the average mass flow rate as

$$W = \rho u A$$ (8.4)

and neglecting the heat generation rate and the viscous dissipation (small as compared to the heat transfer rate), and substituting the definition of fluid enthalpy ($h = e + P/\rho$) into the above equations for a constant area duct, one gets

$$\frac{\partial \rho}{\partial t} + \frac{1}{A}\frac{\partial W}{\partial x} = 0$$ (8.5)

$$\frac{1}{A}\frac{\partial}{\partial t}\left(\frac{W}{\rho}\right) + \frac{W}{\rho A^2}\frac{\partial}{\partial x}\left(\frac{W}{\rho}\right) + \frac{1}{\rho}\frac{\partial p}{\partial x} + g\frac{\partial z}{\partial x} + \frac{f}{di}\frac{W|W|}{2\rho^2 A^2} = 0$$ (8.6)

$$\frac{\partial}{\partial t}(\rho h - P) + \frac{1}{A}\frac{\partial}{\partial x}(Wh) = \frac{\partial q}{\partial y} \tag{8.7}$$

These equations can be applied to determine mass flow rate, pressure, and enthalpy as a function of position and time: $W(x,t)$, $P(x, t)$, and $h(x, t)$, using equations of state which relate the thermodynamic variables. The initial distribution of these variables is known from steady-state calculations. Also needed are the boundary conditions for W, P, h, and the surface heat transfer rate q as a function of position and time.

There are basically three different approaches to obtain a numerical solution for Equations (8.5) to (8.7). They are: (1) the fully compressible flow method (Porshing et al., 1969), (2) the momentum and channel integral method (Agrawal and Guppy, 1978), and (3) the single mass flow rate method (Meyer, 1961). In the fully compressible flow model, the multiple point difference equation approximation to the conservation equations is solved directly for variable W, P, and h using the appropriate equation of state for density:

$$\rho = \rho(h, P) \tag{8.8}$$

and rewriting the mass continuity Equation (8.5) using

$$\frac{\partial \rho}{\partial t} = \left(\frac{\partial \rho}{\partial h}\right)_p \frac{\partial h}{\partial t} + \left(\frac{\partial \rho}{\partial P}\right)_h \frac{\partial P}{\partial t} \tag{8.9}$$

to get

$$\left(\frac{\partial \rho}{\partial h}\right)_p \frac{\partial h}{\partial t} + \left(\frac{\partial \rho}{\partial P}\right)_h \frac{\partial P}{\partial t} + \frac{1}{A}\frac{\partial w}{\partial X} = 0 \tag{8.10}$$

Equation (8.9) can be substituted into the energy Equation (8.7) and along with Equation (8.10) results in a system of equations that can be solved for $\partial P/\partial t$ and $\partial h/\partial t$ to give:

$$\frac{1}{C^2}\frac{\partial P}{\partial t} + \frac{1}{A}\frac{\partial W}{\partial x} + \frac{W}{\rho^2 A}\left(\frac{\partial \rho}{\partial h}\right)_p \frac{\partial P}{\partial x} - \frac{W}{\rho A}\left(\frac{\partial \rho}{\partial h}\right)_p \frac{\partial h}{\partial x} = -\frac{1}{\rho}\left(\frac{\partial \rho}{\partial h}\right)_p \frac{\partial q}{\partial y} \tag{8.11}$$

and

$$\frac{1}{C^2}\frac{\partial \rho}{\partial t} + \frac{1}{\rho A}\frac{\partial W}{\partial x} - \frac{W}{\rho^2 A}\left(\frac{\partial \rho}{\partial P}\right)_h \frac{\partial P}{\partial x} + \frac{W}{\rho A}\left(\frac{\partial \rho}{\partial P}\right)_h \frac{\partial h}{\partial x} = \frac{1}{\rho}\left(\frac{\partial \rho}{\partial P}\right)_h \frac{\partial q}{\partial y} \tag{8.12}$$

where C is the isentropic sonic velocity defined as

$$C = \cfrac{1}{\sqrt{\cfrac{1}{\rho}\left(\cfrac{\partial \rho}{\partial h}\right)_{p} + \left(\cfrac{\partial \rho}{\partial P}\right)_{h}}} \tag{8.13}$$

Equations (8.6), (8.8), (8.11), and (8.12) can be solved for the four unknowns: $W(x, t)$, $\rho(x, t)$, $p(x, t)$, and $h(x, t)$.

The drawback in using this method is its numerical stability (explicit methods) and accuracy (implicit methods), since the required integration time steps are of the order of the time for a sonic wave to pass through one space step; that is,

$$\Delta t \leq \frac{\Delta x}{C + |u|} \tag{8.14}$$

Therefore, it is obvious that the computational time requirements will be prohibitively long for fluid with high sonic velocity. To remove the time step dependency on sonic velocity, one can assume that the fluid density is a function of enthalpy only, that is,

$$\rho = \rho(h,P) \tag{8.15}$$

where P is a spatially constant reference pressure (Meyer, 1961).

This approximation removes the spatial pressure distribution in the flow region and thus simplifies the overall momentum and energy equations, leading to the removal of acoustic wave phenomena. The momentum integral model is limited by numerical stability considerations to integration time steps of the order of the fluid residence time, that is,

$$\Delta t \leq \frac{\Delta x}{|u|} \tag{8.16}$$

This improvement in the integration time step makes this method's applicability to large system studies quite feasible, while retaining the essential physical features of the process for transients in which the duration of significant changes in pressures and velocities are longer than the time for several sonic waves to pass through the system.

Further computational simplification can be achieved by neglecting the mass flow rate distribution along the flow path (single mass flow rate model), that is,

$$\frac{\partial \rho}{\partial t} = -\frac{1}{A}\frac{\partial W}{\partial x} = 0 \tag{8.17}$$

This assumption leads to a considerable simplification in the energy and momentum equations, which allows the determination of temporal variation of mass flow rate through solution of a single ordinary differential equation of the form

$$\frac{1}{A}\frac{dW}{dt} + \frac{W^2}{A^2}\frac{d\left(\frac{1}{\rho}\right)}{dx} + \frac{dp}{dx} + g\rho\frac{dz}{dx} + \frac{f}{di}\frac{W|W|}{2\rho A^2} = 0 \tag{8.18}$$

which may be integrated over a flow length, Δx, using the appropriate pressure boundary conditions.

Numerical comparisons of the momentum integral and single mass flow rate models have shown reasonable agreement (Meyer, 1961). The waterside pressure drop (~5 b) is a small fraction of the operating pressure (~172 b), and hence the effect of spatial and temporal pressure distributions on water/steam properties can be ignored. Hence, the single-mass flow rate method seems to be the most desirable approach for transients.

8.5 MATHEMATICAL MODEL

The mathematical model consists of the energy balance equations for shell, sodium, tube wall, and water. The continuity equation on the water side has been ignored. Momentum equation is not solved, and constant pressure is assumed as mentioned earlier.

Equations (8.19) to (8.22) represent the energy balance equations of Shell, Sodium, water/steam, and tube, respectively.

$$\text{Shell}: \quad C_4(x,t)\frac{\partial T_4}{\partial t}(x,t) = h_{14}(x,t)\left[T_1(x,t) - T_4(x,t)\right] \tag{8.19}$$

$$\text{Sodium}: \quad C_1(x,t)\frac{\partial T_1}{\partial t}(x,t) - F_1(t)Cp_1(x,t)\frac{\partial T_1}{\partial x}(x,t) =$$
$$h_{12}(x,t)\left[T_1(x,t) - T_2(x,t)\right] - h_{14}(x,t)\left[T_1(x,t) - T_4(x,t)\right] \tag{8.20}$$

$$\text{Water}: \quad F_3(t)\frac{\partial H_3}{\partial x}(x,t) + M_w(x,t)\frac{\partial H_3}{\partial t} = h_{23}(x,t)\left[T_2(x,t) - T_3(x,t)\right] \tag{8.21}$$

$$\text{Tube}: \quad C_2(x,t)\frac{\partial T_2}{\partial t}(x,t) = h_{12}(x,t)\left[T_1(x,t) - T_2(x,t)\right]$$
$$- h_{23}(x,t)\left[T_2(x,t) - T_3(x,t)\right] \tag{8.22}$$

where

T_1, T_2, T_3, T_4	Temperature of sodium, tube wall, water, and shell, respectively, °C
C_1, C_2, C_4	Capacitance of sodium, tube wall, and shell, respectively, J/m°C
H_3	Enthalpy of water/steam, J/Kg
M_w	Capacitance of water/steam, Kg/m
Cps	Specific heat of sodium, J/kg°C
F_1, F_3	Sodium and water flow, kg/sec
h_{12}, h_{23}, h_{14}	Product of perimeter and heat transfer coefficient from sodium to metal wall, metal wall to water, and sodium to shell, respectively, W/m°C
P	Water/ steam pressure, bars
x, t	Space and time, respectively

8.6 HEAT TRANSFER CORRELATIONS

The correlations of heat transfer in the different regimes are a function of the prevailing process conditions and geometry. LMFBR steam generators have been of three geometries, namely straight tube, serpentine, and helical. These different geometries have evolved in view of a large temperature difference across the heat transfer length (~200°C), for which the expansion of the tubes and shell needs to be considered to provide requisite flexibility. In the straight vertical design, either the tubes have bends to accommodate the expansion or the shell is provided with bellows. In the serpentine and helical geometries, the flexibility is built in because of geometry.

8.6.1 Single-Phase Liquid Region

The correlations for this regime are of the Dittus-Boelter's type and differ in the range of applicability. However, for coiled tubes, the effect of centrifugal force comes in and must be accounted for. The relations for the different geometries are given in Table 8.2 (Vaidyanathan et al., 2010a).

8.6.2 Nucleate Boiling

Many correlations are available for this regime, of which Roshenow, Thom, Jens and Lottes, Dengler and Adams, and Chen's correlations are important for LMFBR conditions (Collier and Thome, 1994). For many correlations, a complete validity range of pressures, heat flux, void fraction, and mass flow rate are not clear. The heat transfer in this regime is a combination of nucleate boiling and forced convection. Chen (Collier and

Table 8.2 Heat transfer correlations

Fluid	Name	Heat transfer correlation
Sodium	Subbotin (1964)	$Nu = 8\left[\dfrac{d_h}{L} + 0.27\left[\dfrac{E}{d_0} - 1.1\right]^{0.46}\right] Pe^{0.6}$ <div align="right">Straight(SV) and Serpentine(S)</div>
	Dwyer (1963)	$Nu = 1.086\left[5.44 + 0.228Pe^{0.614}\right]\left[\left(\sin\theta + \sin_\theta^2 / (1 + \sin_\theta^2\right)\right]$ <div align="right">Helical(H)</div>
Sub Cooled Water & Steam	Mikheev (1952)	$Nu = 0.021\,Re^{0.8}Pr^{0.43}$ <div align="right">SV and S</div>
	Mori and Nakayama (1967)	$Nu = \dfrac{1}{41}Re^{5/6}\,Pr^{0.4}\left[\dfrac{d_i}{D}\right]^{1/12}\left[1 + \dfrac{0.61}{\left[Re\left[\dfrac{d_i}{D}\right]^{2.5}\right]^{\frac{1}{6}}}\right]$ <div align="right">H</div>
Nucleate boiling	Roshenow and Clark (1951)	$h_b = \left[\dfrac{\mu'r'}{\left(\dfrac{\sigma}{g\left[\rho' - \rho''\right]^{0.5}}\right)}\right]\left[\dfrac{C'}{0.013r'Pr^{1.7}}\right]^3$ $\left(t_w - t_{sat}\right)^3 / \left(t_w - t_f\right)$ $h_c = 0.019\left[\dfrac{k'}{d_i}\right]\left[Gd_i / \mu'\right]^{0.8} Pr'^{0.33}$ <div align="right">SV&S</div>
	Owhadi et al., 1968	$\dfrac{h_{TPF}}{h_L} = A\,exp\left[4.436 - 29.722X_{tt} + 141.237X_{tt}^2 \right.$ $\left. -325.34X_{tt}^3 + 272.58X_{tt}^4\right]$ $h_L = 0.023\left[\dfrac{k'}{d_i}\right]RePr^{0.4}\left[\dfrac{d_i}{D}\right]^{0.1}$ where $A = 1$ when $X_{tt} > 0.05$; $A = \left[\dfrac{D}{20d_i}\right]^{0.25}$ when $X_{tt} \leq 0.05$ <div align="right">H</div>
Critical Quality/ Dryout	Konikov (1966)	$x_{crit} = 76.6\varnothing^{-.125}G^{-.33}(D_i * 10^3)^{-0.07} \,exp$ $(-0.00795P)$ <div align="right">SV&S</div>
	Duchatelle (1973)	$x_{crit} = 1.69 * 10^{-4}\varnothing^{0.719}G^{-0.212} \,exp\,(0.0025P)$ H

(Continued)

Table 8.2 (Continued) Heat transfer correlations

Fluid	Name	Heat transfer correlation	
Post dryout	Miropolsky (1963)	$Nu = Y * 0.023Re^{0.8}Pr_w^{0.43}$	
		$Y = 1 - 0.1\left[\dfrac{\rho'}{\rho''} - 1\right]^{0.4}(1-x)^{0.4}\left[x + \dfrac{\rho''}{\rho'}(1-x)\right]^{0.8}$	SV&S
	Miropolsky and Pikus (Cumo, 1972)	$Nu = C_f Re^{0.8}Pr_w^{0.8}Z_f \quad C_f = 0.017\left[1 + 1.59\dfrac{d_i}{R_{coil}}\right]$	
		$Z_f = Y\left[x + \dfrac{\rho''}{\rho'}(1-x)\right]^{0.8} \quad Y = \dfrac{\rho'}{\rho''}\left[1 - \dfrac{(\rho'-\rho'')}{\rho'}x^{0.2}\right]$	H

Note on nomenclature in the table: H, Helical coil; S, Serpentine; SV, straight vertical.

Thome, 1994) has proposed that the boiling component and the convective component be multiplied by a nucleate boiling suppression factor and Reynolds number correction factor, respectively. However, the factors are validated only up to 36 b pressure. Hence, the Roshenow's correlation (Roshenow and Clark, 1951), which is generally used for pool boiling, has been selected. The convective component is described by the Dittus-Boelter relation. For the helical coil geometry, Owhadi-Bell-Crain experimental data can be used (Owhadi et al., 1968). These correlations are also presented in Table 8.2.

8.6.3 Dry-Out

The point at which dry-out occurs is given by the critical heat flux correlation. The correlation of Konikov and Modnikov (Konikov, 1966) is suitable for straight vertical and serpentine geometries, while Duchatelle's experimental correlation is suited for helical configuration (Duchatelle et al., 1973).

In sodium side correlation: Nusselt (Nu) and Peclet (Pe) numbers are computed from the hydraulic diameter, dh, the liquid metal physical properties are taken at the bulk average temperature, and s, L, and d_O are for the tube pitch, the exchanger length, outside diameter, respectively, and the tube, Θ the helical angle.

For subcooled water/superheated steam correlation, the inside tube diameter d_i is used for calculating Nu and Reynolds (Re) numbers. The water physical properties are taken at the bulk average temperature except for Near wall Prandtl number Pr_w, where the wall temperature is used. D is the diameter of the coil.

For nucleate boiling correlation, the properties taken at saturation temperature, t_{sat}, at the prevailing pressure: μ' dynamic viscosity of water, r' latent heat of vaporization of water, ρ' density of water, ρ'' density of saturated steam, C' = specific heat of water, σ' surface tension of water, k' thermal conductivity of water, Pr Prandtl number of water, C' specific heat of water, g gravity acceleration, G mass flux of two-phase medium, di tube internal diameter, t_w wall temperature, t_f fluid bulk temperature, Xtt Lockhart–Martinelli parameter.

In critical quality correlations, ϕ heat flux, G mass flux, P pressure, and in post dry-out correlation, x is the steam quality

8.6.4 Post Dry-Out

Most of the post dry-out heat transfer correlations are empirical in nature and are related to the saturated vapor heat transfer. Literature survey shows applicability of Miropolsky (1963) correlation for straight vertical and serpentine configurations and Miropolski-Pikus correlation (Cumo, 1972) for helical configuration.

8.6.5 Superheated Region

The correlations for this region are same as those used for single-phase liquid zone.

8.6.6 Sodium Side Heat Transfer

Subbotin (1964) correlation for straight vertical and serpentine geometries and Dwyer (1963) correlation for helical coils are applicable.

8.7 PRESSURE DROP

The total pressure drop consists of the frictional drop, spatial acceleration, and potential drops. For dynamic conditions we need to add the inertial pressure drop also. Friction factor for single-phase flow is given by Colebrook and White correlation (Idelchik, 1966) for straight vertical and serpentine configurations, while Mori-Nakayama correlation (Mori and Nakayama, 1967) is applicable for helical configuration. For a two-phase region, Martinelli-Nelson factor (Martinelli and Nelson, 1948), which is widely reported in literature, can be used. Other correlations are those of Thom, Levy, and Armand (Wallis, 1969), but Martinelli-Nelson factor is used for estimating a two-phase pressure drop in most of the systems. For a bend region, pressure drop is calculated using Hobb's relation (Hobbs, 1961). Correlations used are presented in Table 8.3.

Table 8.3 Pressure drop correlations

	Correlation	Geometry	Reference
Friction Drop	$\Delta P_f = \dfrac{4fLv^2\rho}{2g_c d_i}$ for single phase		
	$\Delta P_f = \left(\dfrac{\Delta P_f}{\Delta X}\right)_{sp} C_{me} L_{tp}$ for two-phase		
	$f = \dfrac{1.0}{\left[-2\log_{10}\left(\dfrac{\varepsilon}{3.7 + 2.51/\mathrm{Re}\sqrt{f}}\right)\right]^2}$	SV and S	
	$f = \left(\dfrac{di}{D}\right)^{0.5} \dfrac{0.192}{\left[R_e\left(\dfrac{di}{D}\right)^{2.5}\right]^{\frac{1}{6}}}\left[1 + \dfrac{0.068}{\left\{\mathrm{Re}\left(\dfrac{di}{D}\right)^{2.5}\right\}^{\frac{1}{6}}}\right]$	H	
Potential Drop	$\Delta P_p = \dfrac{S_i \rho_{av} g}{g_c L}$	SV, S and H	
Acceleration Drop	$\Delta P_a = \dfrac{G^2}{2g_c}\left(\dfrac{1}{\rho_2} - \dfrac{1}{\rho_1}\right)$	SV, S and H	
Bend Frictional Drop	$\Delta P_{bf} = \dfrac{G^2\left(1.6y - \pi R_0 f\right)}{2\rho_{av} g_c}$ Where $R_o = R_b/di$ $Y = Y_1 + Y_2(125f - 2)$ $Y_1 = 0.294043 - 0.2979258 \ln (R_o)$ $\quad + 0.2009208 \ln R_0^2 - 0.010493942 \ln R_0^3$ $Y_2 = 0.2617594 - 0.16735948 R_0$ $\quad + 0.0152060618 R_0^2$	S	Hobbs (1961)
Bend Potential Drop	$\Delta P_{bp} = \dfrac{2R_b \rho_{av} g}{g_c}$	S	

Notes on Nomenclature used:
ΔP_f frictional pressure drop, f fanning's friction factor, L length, v velocity, d_i tube internal diameter, sp single phase, C_{me} Martinelli-Nelson two-phase pressure drop multiplier, tp two phase, ε tube roughness, R_c coil radius, S_i slope of tube, ρ_{av} fluid average density, G mass flux, ρ_1, ρ_2 fluid densities at the beginning and end of mesh, ΔP_{bf} bend frictional pressure drop, R_b bend radius for serpentine geometry, ΔP_{bp} bend potential pressure drop.

8.8 COMPUTATIONAL MODEL

The partial differential equations are converted into algebraic equations by the method of finite differences. The problem then reduces to solving simultaneous nonlinear equations. This problem is tackled by decoupling some of the equations in the time or iteration step. In the present model it is assumed that the sodium temperature change is negligible, and the water side energy equation is carried out first. After this the tube, sodium and shell equations are solved.

8.8.1 Solution of Water/Steam Side Equations

Equation (8.22) written in finite difference form becomes

$$\frac{C_{2ij}}{\Delta_j}\left(T_{2i+1,j+1} - T_{2i+1,j}\right) = Q_{Na} - Q_w \tag{8.23}$$

where subscript 2 refers to wall, Q_{Na} is the heat transferred from sodium, and Q_w is the heat convected by water (i and j refer to space and time mesh indices, respectively), and further,

$$Q_{Na} = h_{12ij}\left(T_{1i+1,j+1} - T_{2i+1,j+1}\right) \tag{8.24}$$

$$Q_w = h_{23i+1,j+1}\left(T_{2i+1,j+1} - T_{3i+1,j+1}\right) \tag{8.25}$$

where subscripts 1 and 3 refer to sodium and water, respectively.
From Equation (8.25),

$$T_{2i+1,j+1} = \frac{Q_w}{h_{23ij}} + T_{3i+1,j+1}$$

Substituting this in Equation (8.24) (i.e., eliminating $T_{2i+1,j+1}$),

$$Q_{Na} = h_{12ij}\left(T_{1i+1,j+1} - \frac{Q_w}{h_{23ij}} - T_{3i+1,j+1}\right) \tag{8.26}$$

Substituting for Q_{Na} and $T^{2i+1j+1}$ in Equation (8.23) and rearranging, we get

$$Q_w = \frac{h_{12ij}T_{1i+1,j+1} - T_{3i+1,j+1}\left(h_{23ij} + \dfrac{C_{2ij}}{\Delta_j}\right) + T_{2i+1j}\dfrac{C_{2ij}}{\Delta_j}}{\left(\dfrac{C_{2ij}}{\Delta_j h_{23ij}} + \dfrac{h_{12ij}}{h_{23ij}} + 1\right)} \tag{8.27}$$

The enthalpy Equation (8.21) written in finite difference form becomes

$$\frac{M_{\omega ij}}{\Delta_j}\left(H_{3i+1j+1} - H_{3i+1j}\right) = Q_w - \frac{F_{3ij}}{\Delta_i}\left(H_{3i+1,j+1} - H_{3ij+1}\right) \tag{8.28}$$

Substituting for Q_ω from Equation (8.27) and using the substitution
$T_{3i+1j+1} = T_{3i+1j} + \dfrac{H_{3i+1j+1} - H_{3i+1j}}{C_{pij}}$ into Equation (8.28) we get

$$H_{3i+1j+1} = \frac{\left[\dfrac{M_{wij}}{\Delta_j} H_{3i+1j} + \dfrac{F_{3ij}}{\Delta_i} H_{3ij+1}\right]}{\left[\dfrac{M_{wij}}{\Delta_j} + \dfrac{F_{3ij}}{\Delta_i}\right]\left[\dfrac{c_{2ij}}{\Delta_{jh23ij}} + \dfrac{h_{12ij}}{h_{23ij}} + 1\right]} \tag{8.29}$$

Equation (8.29) is solved for all mesh points from the cold end of the steam generator for a particular time step using the approximation $T_{1i+1,\,j+1} = T_{1i+1,j}$

8.8.2 Solution of Sodium, Shell, and Tube Wall Equations

Equations (8.19) to (8.22) written in finite difference form gives

$$\frac{C_{4ij}}{\Delta_j}\left(T_{4ij+1} - T_{4ij}\right) = h_{14ij}\left(T_{1ij+1} - T_{4ij+1}\right) \tag{8.30}$$

where subscript 4 refers to shell.

$$\frac{C_{2ij}}{\Delta_j}\left(T_{2ij+1} - T_{2ij}\right) = h_{12ij}\left(T_{1ij+1} - T_{2ij+1}\right) - h_{23ij}\left(T_{2ij+1} - T_{3ij+1}\right) \tag{8.31}$$

$$\frac{C_{1ij}}{\Delta_j}\left(T_{1ij+1} - T_{1ij}\right) = \frac{F_{ij}C_{ps}}{\Delta_i}\left(T_{1i+1j+1} - T_{1ij+1}\right) - h_{12ij}\left(T_{1ij+1} - T_{2ij+1}\right)$$
$$-h_{14ij}\left(T_{1ij+1} - T_{4ij+1}\right) \tag{8.32}$$

The unknowns in the above three equations are T_{1ij+1}, T_{2ij+1}, and, T_{4ij+1}. They can be solved using the Gaussian elimination technique or by eliminating the individual variables from the equations.

From Equation (8.30),

$$T_{4ij+1} = \frac{h_{14ij}T_{1ij+1} + \dfrac{c_{4ij}}{\Delta_j} T_{4ij}}{\dfrac{C_{4ij}}{\Delta_j} + h_{14ij}} \tag{8.33}$$

From Equation (8.31),

$$T_{2ij+1} = \frac{h_{12ij}T_{1ij+1} + \dfrac{c_{2ij}}{\Delta_j}T_{2ij} + h_{23ij}T_{3ij+1}}{\dfrac{C_{2ij}}{\Delta_j} + h_{12ij} + h_{23ij}} \tag{8.34}$$

Substituting for T^{2ij+1} and T^{4ij+1} in Equation (8.32) above, we get, after simplifying,

$$T_{1ij+1} = \frac{\psi 5}{\psi 4} \tag{8.35}$$

where

$$\psi 4 = \frac{C_{1ij}}{\Delta_j} + \frac{F_{1j}C_{ps}}{\Delta_j} + h_{12ij} + h_{14ij} - \frac{h^2_{12ij}}{\psi 1} - \frac{h^2_{14ij}}{\psi 2}$$

$$\psi 5 = \frac{C_{1ij}}{\Delta_j}T_{1ij} + \frac{F_{1j}C_{ps}}{\Delta_i}T_{1i+1j+1} + h_{12ij}\frac{\psi 3}{\psi 1} + \frac{h_{14ij}}{\psi 2}\frac{C_{4ij}}{\Delta_j}T_{4ij}$$

$$\psi 1 = \frac{C_{2ij}}{\Delta_j} + h_{12ij} + h_{23ij}, \qquad \psi 2 = \frac{C_{4ij}}{\Delta_j} + h_{14ij},$$

$$\psi 3 = \frac{C_{2ij}}{\Delta_j}T_{2ij} + h_{23ij} + h_{3ij+1}$$

Equation (8.35) is solved for $T_{1i+1j+1}$ for all meshes from the hot end of the steam generator for a particular time step after solving the enthalpy equation of the water/steam side. Knowing T_{1ij+1} we can determine T_{4ij+1} and T_{2ij+1} from Equations (8.33) and (8.34), respectively.

8.9 STEAM GENERATOR MODEL VALIDATION

To assess the applicability of the different heat transfer and pressure drop correlations used, SG sizing was carried out for different designs, the design data of which are reported in open literature (Vaidyanathan et al., 2010b). The predicted design results and the design reported in literature are also given in Table 8.4. The code can predict the overall heat transfer length with a good degree of accuracy in all the designs reported. Deviation is within ±5% of overall heat transfer area. Differences exist in the thermal conductivity value for the materials used for SG in fast reactors. Variations in thermal conductivity up to 15% are reported in ferritic steels in France.

Table 8.4 Comparison of DESOPT code with reported designs

Output data	PHENIX (Serpentine) (Duchatelle et al., 1974)		SNR 300 (Straight vertical) (De Clerq and Van Waveren, 1970)		FBTR (Serpentine) (Srinivasan et al., 2006)	
	CODE	Design	CODE	Design	CODE	Design
Surface area, m²	35.7	37.42	209.9	221.1	271.2	267.4
Water side ΔP, bars	7.13	8.0	2.33	2.6	5.08	4.97
Economizer length, m	25.14	23.76	9.52	10.0	32.24	32.91
Evaporator length, m	27.20	31.89	8.89	9.4	38.64	37.68
Sup heater length, m	5.49	5.11	-	-	20.59	19.61
Ht. transfer length, m	57.83	60.76	18.41	19.4	91.47	90.2

To validate the model a steam generator test facility (SGTF), a model SG rated for 5.5 MW has been set up at Kalpakkam, India. During the full-power operation, by keeping all the process parameters at rated value, it was possible to achieve the rated power for a sodium inlet temperature of 518°C compared to design value of 525°C. This is because of the extra margins provided in the heat transfer area. The SGTF is equipped with thermocouples on the outer side of the tubes, thus measuring the outer wall temperature, which should be close to the sodium temperature considering a ~ 15°C drop in sodium film.

With the experimentally obtained parameters during the SG power operation, performance of SG was predicted using the code, and the temperature profile in SG thus obtained is shown in Figure 8.9 (Vaidyanathan et al., 2010b).

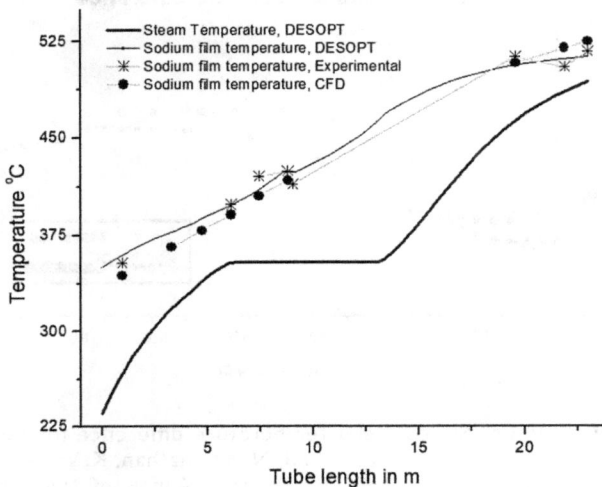

Figure 8.9 Steady-state temperature predictions and SGTF measurements.

Also indicated are experimentally measured sodium temperatures across the height of SG at some locations. The sodium temperatures measured lie quite close (considering thermocouple contact resistance and measurement errors) to the predicted curve for sodium temperature. Also plotted are the 3D calculations of sodium side of steam generator based on CFD analysis (Nandakumar et al., 2012). This gives credence to the predictive capability of the 1D model.

The validation of the transient modeling has been carried out based on tests in the fast breeder test reactor during commissioning (Vaidyanathan et.al., 2010a). The event is the loss of off-site power. The primary and secondary pump drives are provided with flywheels, and hence the coast-down is gradual. The boiler feed pump (BFP) does not have any inertia in the drive, and hence the water flow comes almost to zero instantly. Diesel generators start in 10 s and supply power to maintain the primary and secondary pumps at 300 rpm. The reactor is tripped based on the power failure signal. Due to power failure, the water flow to the steam generator is lost almost instantaneously, but secondary sodium flow circulation is maintained for quite some time due to drive and hydraulic inertias. As a result, temperature of secondary sodium at the SG outlet increases, leading to an increase in temperature difference between the water inlet and the secondary sodium outlet. The increase is from 100°C to 225°C in about 5 min. After ~10 min, the difference starts decreasing due to the fall in sodium temperatures. The measured and predicted values of this temperature difference are compared in Figure 8.10. The comparison is reasonable. Thus, the steady-state and transient modeling of the steam generator along with the indicated heat transfer correlations stands validated.

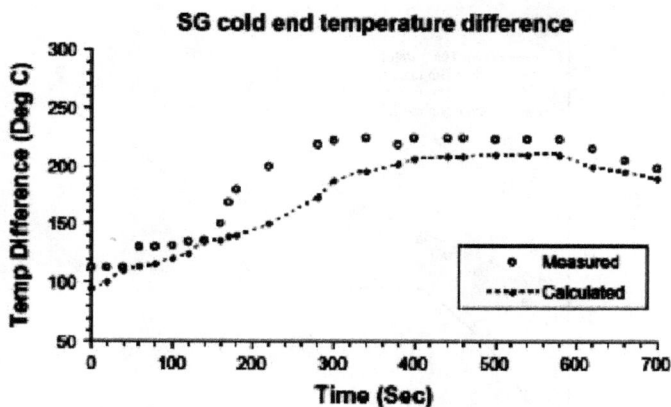

Figure 8.10 Evolution of SG cold end temperature difference for loss of offsite power event. (G. Vaidyanathan, N. Kasinathan, K. Velusamy, Dynamic Model of Fast Breeder Test Reactor. Annals of Nuclear Energy 37, 2010.)

ASSIGNMENT

1. What is the difference between a drum type and a once-through steam generator or boiler? Why is it that most of the SFRs use once-through type steam generators?
2. What are the different heat transfer regimes encountered in a once-through steam generator as the water enters the steam generator and comes out as superheated steam?
3. What is the effect on temperatures in the tubes when the heat transfer mode changes from nucleate boiling to post dry-out?

REFERENCES

Agrawal A.K. and Guppy J. (1978), "An advanced thermohydraulic simulation code for transients in LMFBRs", Brookhaven National Laboratory, BNL-NUREG-50773.

Chetal S.C. and Vaidyanathan G., (1997), "Evolution of design of steam generators for sodium cooled fast reactors", *Proc. Third Conf. on Heat Exchanger Design, Boiler & Pressure Vessels*, Alexandria, Egypt, April.

Collier, J.G. and Thome J.R. (1994), "*Convective Boiling and Condensation*", Oxford University Press, 3rd Edition.

Cumo M. (1972), "Influence of curvature in post dryout heat transfer", *Int. J. Heat Mass Transfer*, Vol. 15, pp. 2045–2062.

De Clerq, W.J.C. and Van Waveren, N.J. (1970), "Steam generator and intermediate heat exchanger development", Proc. Sodium Cooled Fast Reactor Engg., Monaco, pp. 433–453.

Duchatelle L., Nucheze L. and Robin M.G. (1973), "Departure from nucleate boiling in helical tubes of liquid metal heated steam generators", ASME paper 73-HT-57.

Duchatelle L., Nucheze L. and Robin M.G. (1974), "Theoretical and experimental study of PHENIX steam generator prototype modules", *Nucl. Technol.*, Vol. 24, p. 123.

Dwyer O.E. (1963), "Eddy transport in liquid metal heat transfer", *AIChE J.*, Vol. 9, p. 2.

Hobbs J.L. (1961), "Pressure loss computations on incompressible fluid flow", General Electric Report, APEX-754.

Idelchik I.E. (1966), "Handbook of Hydraulic Resistances", AEC-TR-6630.

International Atomic Energy Agency, (1990), "International Working Group on Fast Reactors", Proceedings of the Specialists' Meeting On Acoustic/Ultrasonic Detection of In Sodium Water Leaks On Steam Generators, IWGFR/79, Aix-En-Provence, France, October 1990.

Konikov A.S. (1966), "Experimental study of the conditions under which heat exchange deteriorates when steam water mixture flows in heated tubes", *Teploenergetika*, Vol. 12, p. 13.

Martinelli R.C. and Nelson D.B. (1948), "Prediction of pressure drop during forced circulation of boiling water", Trans. ASME, 70.

Meyer J.E. (1961), "Hydrodynamic models for the treatment of reactor thermal transients", *Nuc. Sci. Engg.*, Vol. 10, p. 269.

Mikheev M.A. (1952), "Convection forcee en ecoulement longitudinal", *Izv.Acad. Nauk, Tranduction CNRS 335*, Vol. 10, p. 1448.

Miropolsky Z.L. (1963), "Heat transfer in film boiling of a steam water mixture in steam generating tubes", *Teploenergetika*, Vol. 10, No. 5, pp. 49–52; transl. AEC-TR-6252, 1964.

Mori Y. and Nakayama, W. (1967), "Study on forced convective heat transfer in curved pipes", *Int. J. Heat and Mass Transfer*, Vol. 10, pp. 681–695

Nandakumar R, Selvaraj P, Athmalingam S, Balasubramaniyan V, Chetal S.C. (2012), "Thermal simulation of sodium heated once through steam generator for a fast reactor", *Int. J. Adv. Eng. Sci. Appl. Math.*, July–September, Vol. 4, No. 3, pp. 127–137. https://doi.org/10.1007/s12572-012-0063-1

Owhadi, A., Crain, B. and Bell, K.J. (1968), "Forced convection boiling inside helically coiled tubes", *Int. J. Heat and Mass Transfer*, Vol. 11, pp. 1779–1793.

Porshing T.A., Murphy J.H., Redfield J.A. and Davis V.C. (1969), "FLASH-4: A fully implicit Fortran IV program for digital simulation of transients in a reactor plant", WARD-TM-840.

RELAP3B (1974), "A reactor system transient code", Brookhaven National Lab., RP-1035.

Roshenow W.M. and Clark J.A. (1951), "Heat transfer and pressure drop data for high flux densities to water at high sub critical pressures", Heat Transfer and Fluid Mechanics Inst.

Srinivasan G., Suresh Kumar K.V., Rajendran B. and Ramalingam P.V. (2006), "Fast breeder test reactor-design and operating experience", *Nucl. Eng. Des.*, Vol. 236, p. 796; https://doi.org/10.1016/j.nucengdes.2005.09.024

Subbotin V.I. (1964), "Heat removal from reactor fuel elements cooled by liquid metals", *Proc. Third Int. Conf. on Peaceful uses of Atomic Energy*, Conf-28/ P3288, Geneva.

Vaidyanathan G., Kasinathan N. and Velusamy K. (2010a), "Dynamic model of fast breeder test reactor", *Annals of Nuclear Energy*, Vol. 37, pp. 450–462. https://doi.org/10.1016/j.anucene.2010.01.013.

Vaidyanathan G., Kothandaraman A.L., Siva Kumar L.S., Vinod V., Noushad I.B., Rajan K.K. and Kalyanasundaram P. (2010b), "Development of one-dimensional computer code DESOPT for thermal hydraulic design of sodium-heated once through steam generators, December 2010", *Int. J. Nucl. Energy Sci. Technol.*, Vol. 5, No. 2. https://doi.org/10.1504/IJNEST.2010.030556.

Vaidyanathan, G. (2013), *Nuclear Reactor Engineering—Principles and Concepts*, S.Chand Publishing, Delhi, India.

Wallis G.B. (1969), *One-Dimensional Two-Phase Flow*, Mc Graw Hill Book Company, New York.

Chapter 9

Computer Code Development

9.1 INTRODUCTION

DYNAM is a one-dimensional transient analysis system code developed based on the mathematical models described in the previous chapters (Vaidyanathan et al., 2010). It represents the two loops of FBTR plant from reactor to steam generator. Schematic representation of reactor and one loop is depicted in Figure 9.1. DYNAM however models both loops as many analyses involve effect of event in one loop on the other. Since the primary sodium system contains two loops which are hydraulically linked with the core, the model for primary hydraulics includes both the loops. DYANA-P (Natesan et al., 2012) is a one-dimensional system dynamics code developed for performing plant dynamics studies for Prototype Fast Breeder reactor (PFBR). It uses the various component models from DYNAM code. STITH-2D is a two-dimensional computational fluid dynamics (CFD) code used to carry out a 2-dimensional analysis of hot pool of PFBR. DYANA-HM uses the modules of DYANA-P except for the hot pool which is modelled with STITH-2D (Natesan and Velusamy, 2019). This chapter presents a description of the code modules of DYNAM and goes on to present the comparison of event analysis for PFBR with DYANA-P and DYANA-HM and discuss the results.

9.2 ORGANIZATION OF DYNAM

DYNAM code is organized in a modular fashion. It comprises the following major programs:

MAIN – Serves as the main link between different programs and decides the flow of the calculations. All input data are read here, and output results printed as per desired format and frequency.

CORE – Calculates reactor neutronic and thermal power and temperatures in the different part of the core based on flow inputs from PRYHYD program.

DOI: 10.1201/9781003283188-9

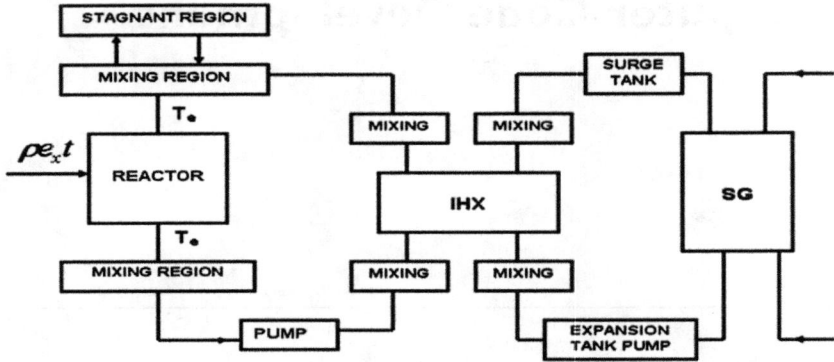

Figure 9.1 Schematic of FBTR model for dynamic analysis.

PRYHYD – Calculates the sodium flow in the different parts of the primary sodium loop including flow distribution in the core. It also calculates the levels in the reactor and various capacities.

SECHYD – Calculates sodium flow in the secondary loops

IHX – Calculates the temperature distribution in primary and secondary in the IHX for a given primary flow, primary inlet temperature, secondary flow, and secondary inlet temperature.

PILOSS – calculates piping heat losses from primary and secondary sodium systems.

SGEN – Calculates the temperature distribution on secondary sodium and water/steam in SG for given water flow, secondary sodium flow, secondary sodium, and water inlet temperature to SG.

The logic flow of the DYNAM code is given in Figure 9.2. First the code reads the geometrical and other data required for steady state calculations. In the steady state the calculations start from steam generator and proceed towards IHX after calculation of heat losses in secondary piping. The IHX calculations are then carried out and after the calculations of piping losses in primary, the CORE thermal model and PRYHYD models are solved. For transient studies, the code can handle perturbations in:

i) Reactivity
ii) Primary pump speed in both or one loop
iii) Secondary pump speed in both or one loop
iv) Water flow
v) SG outlet pressure

For the transient calculations, the hydraulics of the primary and secondary of both loops are solved for a given time step. The solution of the reactor kinetics equations is carried out next giving the power and reactivity

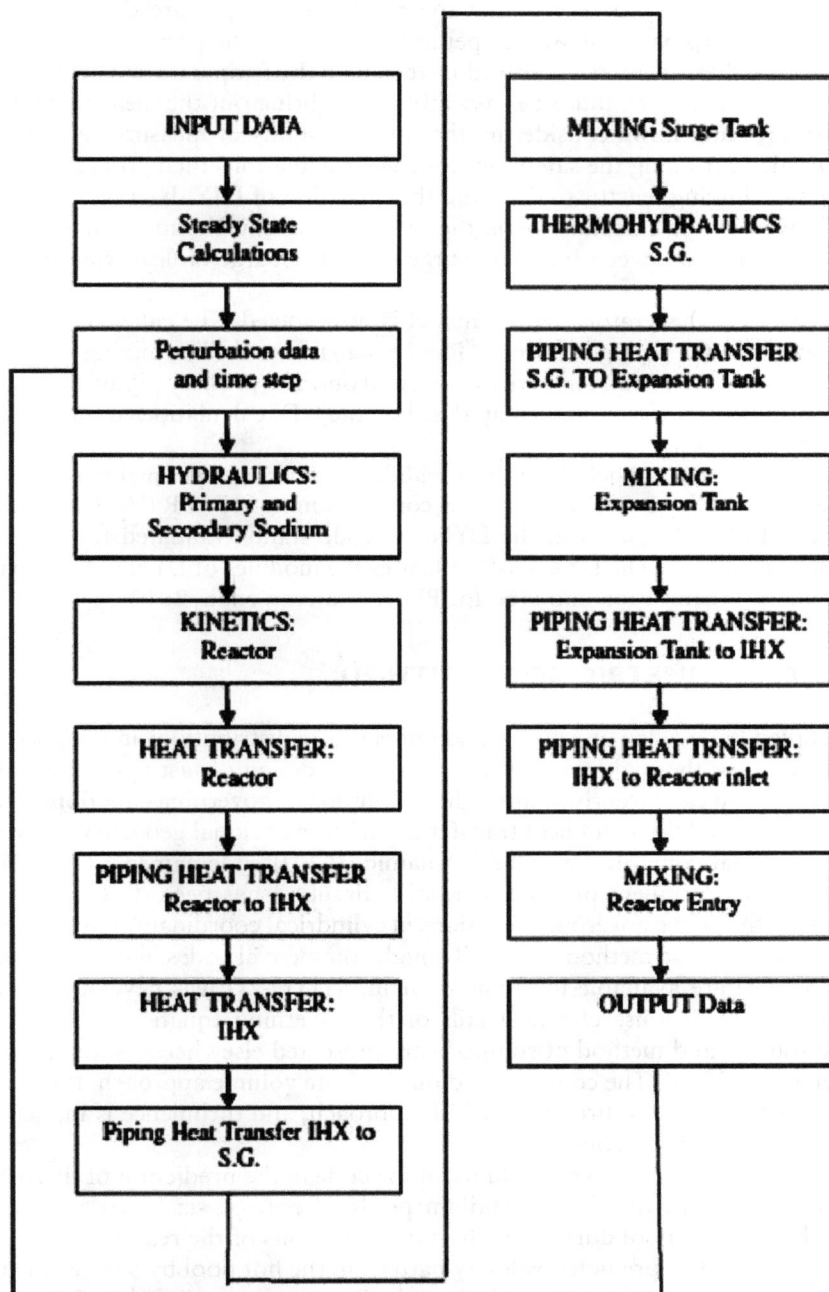

Figure 9.2 Organization of DYNAM code.

feedbacks. It may be noted that the reactivity changes are calculated in each time step based on the temperature changes in the previous time step. The core thermal model is solved next to give the temperature distribution in the core, blanket and other assemblies and bring out the measured temperatures and flows considering the time constants of measurement. This is needed for fixing the safety settings. The calculations then proceed along the pipe linking reactor to IHX and then solution of IHX dynamics. This is followed by the calculations on the secondary circuit thermal calculations on the piping between IHX and surge tank. The surge tank mixing model is solved next.

The steam generator thermal model is now solved. The calculation then proceeds to the piping between SG and expansion tank. The mixing calculations in the expansion tank are then carried out. The primary piping between IHX and reactor inlet is taken up based on the IHX calculations done earlier in the time step.

The different models have been validated against experiments in other reactors and those carried out during commissioning in FBTR (Vaidyanathan et al., 1994, 2010). Thus, the DYNAM code stands validated for system transient studies. The DYANA-P code uses the modules of DYNAM and has a similar organization and used for PFBR (Natesan et al., 2012).

9.3 AXISYMMETRIC CODE STITH-2D

Detailed knowledge of velocity and temperature distributions in the pool is essential for thermal hydraulic and structural design of fast reactors. Such information can be derived through the solution of governing equations representing fluid flow and heat transfer in multidimensional geometry. A two-dimensional computational fluid dynamics (CFD) code-named STITH-2D (structured turbulent induced thermal hydraulics) has been developed for solving the above governing equations in cylindrical coordinate system using the finite volume method (FVM). Though commercial codes like Fluent and STAR-CD are available in public domain, STITH-2D was developed as a proprietary in-house effort. Details of the governing equations, their discretization, and method of solutions are presented elsewhere (Natesan and Velusamy, 2019). The code is based on the finite volume approach. Pressure velocity coupling is through SIMPLE approach, and turbulence is modeled by a standard K–€ model.

To demonstrate the applicability of the code in the prediction of thermal hydraulic behavior of large sodium pools of reactor size, axisymmetric analysis of hot pool during steady-state conditions of the reactor has been carried out. The predicted velocity pattern in the hot pool by STITH-2D is shown in Figure 9.3 along with the velocity pattern predicted by Star-CD 3.26 code. It can be observed that both predictions are very close. The nature of flow in the region above the core as well as near the inner vessel

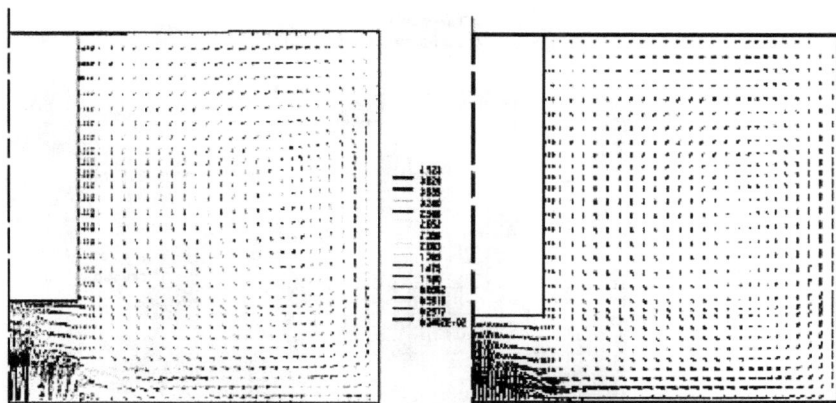

Figure 9.3 Comparison of hot pool velocity patterns between STAR-CD and STITH-2D.

Source: Natesan K., Velusamy K., Coupled System Dynamics and Computational Fluid Dynamics Simulation of Plant Transients in Sodium Cooled Fast Reactors. *Nuclear Engineering and Design* 342, 2019.

is predicted similarly by both codes. The predictions of a recirculating flow pattern in the hot pool are also similar. This is a validation of the STITH-2D code.

9.4 ONE-DIMENSIONAL CFD-COUPLED DYNAMICS TOOL

A two-dimensional axisymmetric model of hot pool has been formulated using the STITH-2D code, while the rest of the plant, namely core, cold pool, intermediate heat exchangers, secondary sodium circuit, piping, and steam generators, is modeled using the one-dimensional formulations available in the plant dynamics code DYANA-P (Natesan and Velusamy, 2019).

The coupled code system is named DYANA-HM (with hot pool modeling). Modeling of the hot pool is done in STITH-2D code as shown in Figure 9.4, except that the inclined portion of the inner vessel is modeled as flat at the location of core top. This is to simplify the model and thereby to save computational time. Sodium flow and outlet temperature distribution in various radial zones of reactor core, obtained from the one-dimensional multi-zone modeling of core (from DYANA-P), is specified as inlet boundary condition to the 2D axisymmetric model of hot pool. Outflow from the hot pool is modeled through mass sink specified at the location of inlet window of IHX in the STITH-2D model. IHX inlet temperature required for the system model (1D model) is evaluated as the average temperature of fluid

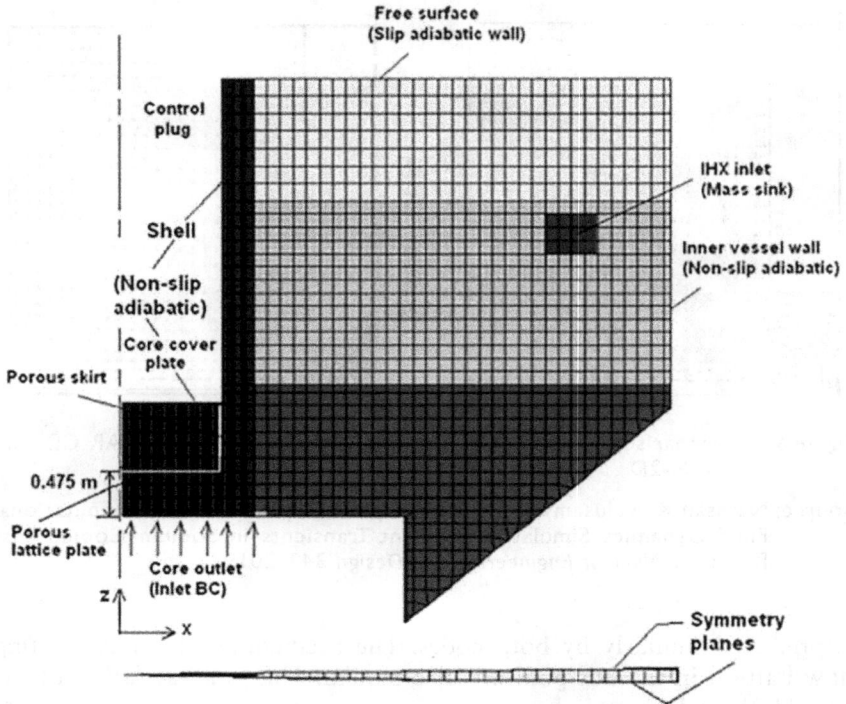

Figure 9.4 Axisymmetric model of PFBR hot pool.

Source: Natesan K., Velusamy K., Coupled System Dynamics and Computational
      Fluid Dynamics Simulation of Plant Transients in Sodium Cooled Fast
      Reactors. *Nuclear Engineering and Design* 342, 2019.

entering the IHX inlet window (mass sink) in the 2D axisymmetric model of
the hot pool.

9.5 COMPARISON OF PREDICTIONS OF DYANA-P AND DYANA-HM

Four different typical plant transients, namely (1) spurious SCRAM, (2)
loss of steam water system, (3) class IV power failure, and (4) station blackout events, have been analyzed with both codes for PFBR (Natesan and
Velusamy, 2019). Spurious SCRAM is characterized by reactor SCRAM at
t = 0 s, coast-down of primary and secondary pumps to 20% speeds during t = 0 to 100 s, and feed water flow to SG reduced to 15% (during t = 0
to 30 s) gradually by the SG sodium outlet temperature controller acting
on water flow to SG. Loss of the steam water system is characterized by
total loss of feed water flow to SG at t = 0 s, reactor SCRAM at t = 30 s,

and coast-down of primary and secondary pumps to 20% speeds during t = 30 to 130 s. Class IV power failure is characterized by coast-down of primary sodium pumps to 18%, trip of secondary sodium pumps and total loss of feed water flow to SG at t = 0 s, and reactor SCRAM at t = 0.5 s. Station blackout is simulated by considering that primary and secondary sodium pumps get tripped simultaneously and coast down to zero speed. Heat sink to all of the SG is also lost instantaneously. Due to the speed reduction of primary sodium pumps, the reactor trips automatically by the action of the plant protection system. Accordingly, reactor SCRAM is considered to take place at 0.5 s in this analysis. Sodium flows in both primary and secondary sodium circuits enter natural convection conditions during this event.

Figures 9.5 to 9.7 compare the IHX primary and secondary temperatures evolution with both codes for the above events. From Figure 9.5 for the event of spurious SCRAM, it can be observed that due to the modeling of thermal stratification in DYANA-HM, the IHX primary inlet temperature reduction is gradual during the initial period compared to that predicted by DYANA-P. Significant difference between the predictions exists up to 400 s. However, predictions of primary sodium temperature at the outlet of IHX by both codes are the same.

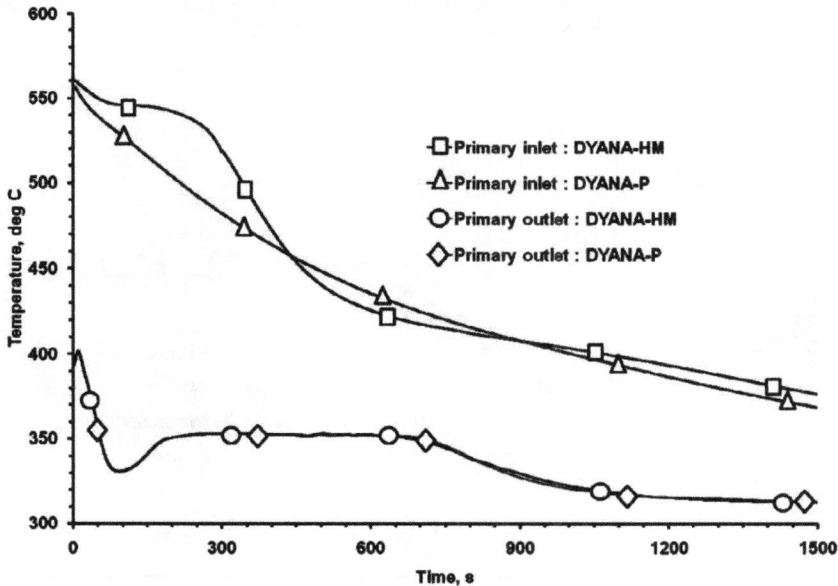

Figure 9.5 IHX primary temperatures evolution—spurious SCRAM.

Source: Natesan K., Velusamy K., Coupled System Dynamics and Computational Fluid Dynamics Simulation of Plant Transients in Sodium Cooled Fast Reactors. *Nuclear Engineering and Design* 342, 2019.

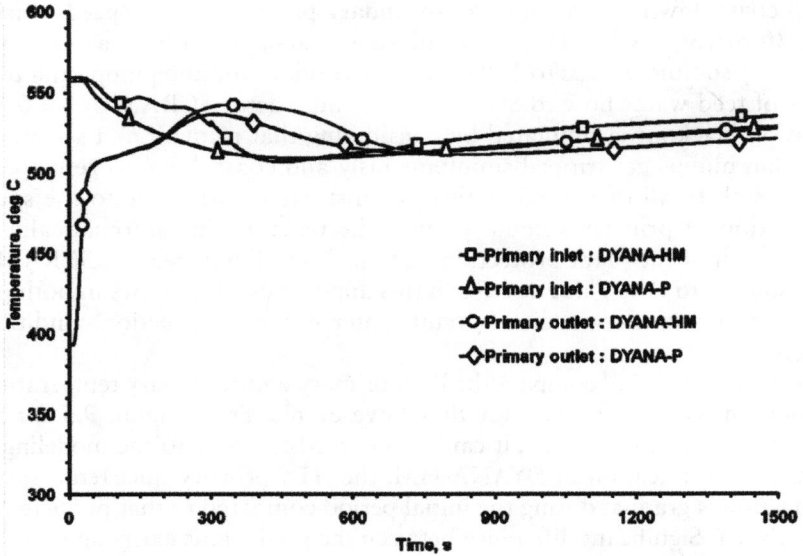

Figure 9.6 IHX primary temperatures—loss of steam water system.

Source: Natesan K., Velusamy K., Coupled System Dynamics and Computational Fluid Dynamics Simulation of Plant Transients in Sodium Cooled Fast Reactors. *Nuclear Engineering and Design* 342, 2019.

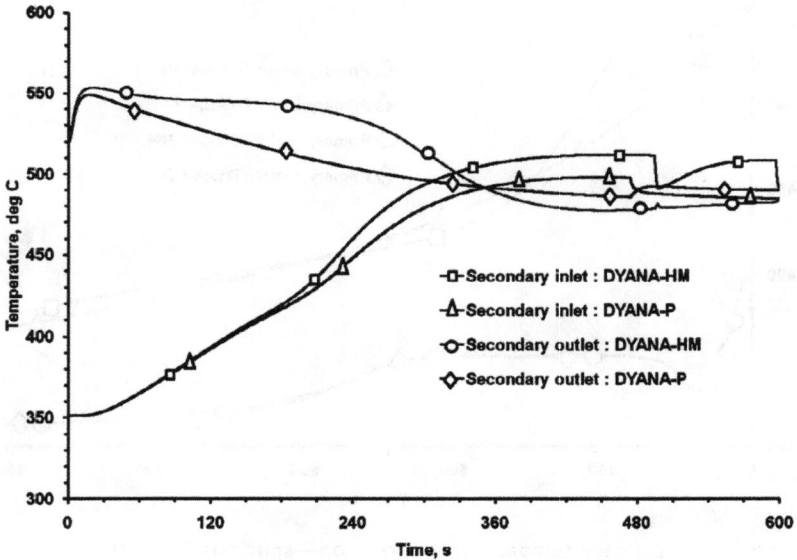

Figure 9.7 IHX primary temperatures—class IV power failure.

Source: Natesan K., Velusamy K., Coupled System Dynamics and Computational Fluid Dynamics Simulation of Plant Transients in Sodium Cooled Fast Reactors. *Nuclear Engineering and Design* 342, 2019.

From Figure 9.6 it can be observed that significant differences exist between the predictions in the hot leg of IHX. The differences between the predictions continue to be present throughout the transient. Predicted temperature evolutions at the outlet of IHX by both codes are close during the first 200 s of the transient. Subsequently, the effect of thermal stratification in the hot pool gets propagated to the outlet of IHX after some time delay. Predicted temperature evolutions at the outlet of IHX by both codes are close.

Evolution of the IHX temperatures for the event of class IV power failure is presented in Figure 9.7. It can be observed that significant differences exist between the predictions in the hot leg temperature of IHX. DYANA-HM predicted hot leg primary temperature falls slower then DYANA-P essentially due to thermal stratification in the hot pool.

The natural convection flow development for the same event is presented in Figure 9.8. After about 300 s, the IHX primary sodium outlet temperature decreases below the inlet temperature (Figure 9.7), indicating heat transfer from the secondary sodium circuit to the primary one. This causes an adverse buoyancy head to develop in the secondary sodium circuit, resulting in flow reversal and flow oscillations. The first evidence of flow reversal in the secondary sodium circuit is predicted by DYANA-P code at

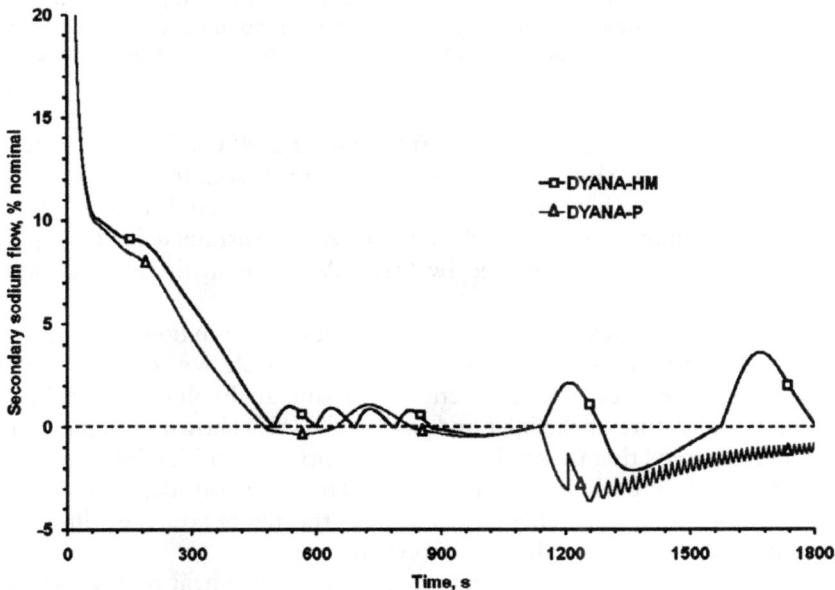

Figure 9.8 Secondary natural convection flow—class IV power failure.

Source: Natesan K., Velusamy K., Coupled System Dynamics and Computational Fluid Dynamics Simulation of Plant Transients in Sodium Cooled Fast Reactors. *Nuclear Engineering and Design* 342, 2019.

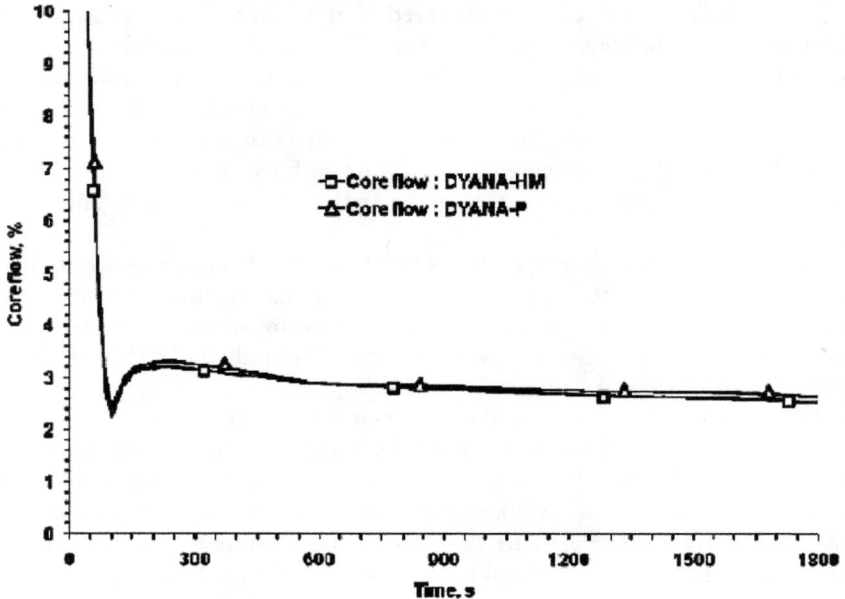

Figure 9.9 Evolution of core flows—station blackout.

Source: K. Natesan, Development of mathematical models and investigation of FBR plant behaviour during transients by a coupled single and multi-dimensional approach, Doctoral Thesis, Homi Bhabha National University, May 2020.

470 s, and the same is predicted by DYANA-HM at 490 s. This difference is due to the relatively slower reduction in the primary sodium temperature at the inlet of IHX due to thermal stratification in the hot pool. The reversal of secondary sodium flow predicted by DYANA-P is sustained for the longer duration, whereas that predicted by DYANA-HM is in the form of short time pulses.

For the station blackout event, the evolution of sodium flows through the core predicted by DYANA-P and DYANA-HM are shown in Figure 9.9. It can be observed that both predictions are similar. Evolution of primary sodium temperature at the inlet and outlet of IHX is shown in Figure 9.10. It can be observed that the evolutions of IHX primary sodium inlet temperature predicted by both models are similar. Thus, the consideration of thermal stratification effect does not influence the temperature evolution of various parts of the plant during this event.

Investigation of typical transients suggest that when heat removal is continued in steam generator units, the prediction of hot leg temperatures of heat exchangers alone is affected due to hot pool thermal stratification. Prediction of cold leg temperatures, and hence the core temperatures under

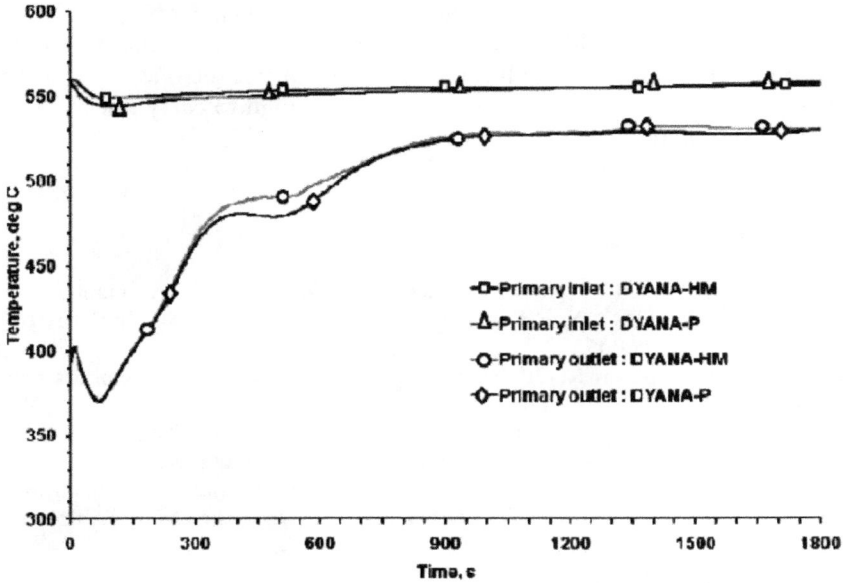

Figure 9.10 Evolution of IHX primary temperatures-station blackout.

Source: K. Natesan, Development of mathematical models and investigation of FBR plant behaviour during transients by a coupled single and multi-dimensional approach, Doctoral Thesis, Homi Bhabha National University, May 2020.

this class of transients, is not affected. However, when no heat removal takes place through SG, hot pool thermal stratification significantly influences temperature evolution in various parts of the plant. This occurs even when the secondary sodium circuit is under forced flow conditions. During transients in which the secondary sodium circuit is under natural convection and the primary circuit is under forced flow, the evolution of natural convection flow behavior in the secondary sodium circuit is influenced by the thermal stratification.

Hot pool is in well-mixed condition during normal operating conditions, and thermal stratification effects develop only after reactor SCRAM. Therefore, the actuation time of the plant protection system during various design basis events does not get affected due to stratification effects. Therefore, multidimensional model-based studies are not essential for the design of the plant protection system and demonstration of core safety. However, multidimensional model-based studies provide accurate information with respect to transient thermal loading on various structures and components that form an essential part of the thermomechanical design and life prediction of systems and components. Moreover, realistic prediction of the transient scenario in the plant under post-shutdown cooling conditions can be obtained only with the inclusion of multidimensional thermal stratification models. This chapter has brought out the fact the one-dimensional codes like DYNAM and DYANA-P are good enough to carry out dynamic simulation of plant transients.

REFERENCES

Natesan K., et al., (2012), "Plant dynamics studies towards design of plant protection system for PFBR", *Nuc. Eng. Design*, Vol. 250, pp. 339– 350; doi:10.1016/j.nucengdes.2012.05.009.

Natesan K. and Velusamy K. (2019), "Coupled system dynamics and computational fluid dynamics simulation of plant transients in sodium cooled fast reactors", *Nuc. Eng. Des*, Vol. 342, pp. 157–169; doi:10.1016/j.nucengdes.2018.12.001.

Natesan K. (2020), Development of mathematical models and investigation of FBR plant behaviour during transients by a coupled single and multi-dimensional approach, Doctoral Thesis, Homi Bhabha National University, May, 2020; http://www.hbni.ac.in/phdthesis/engg/ENGG02201304010.pdf

Vaidyanathan, G. et al., (1994), Dynamic tests related to undercooling events in FBTR, in: Proceedings of the Int. Top. Meet. on Sodium Cooled Fast Reactor Safety, vol. 1, Obninsk, Russia, pp 156–165.

Vaidyanathan G., Kasinathan N., Velusamy K., (2010), "Dynamic model of fast breeder test reactor", *Ann. Nucl. Energy*, Vol. 37, pp. 450–462; doi:10.1016/j.anucene.2010.01.013.

Chapter 10

Specifying Sodium Pumps Coast-Down Time

10.1 INTRODUCTION

Heat generated in a reactor is removed by primary sodium. During transients resulting from loss of pumping power, coolant flow through the reactor core reduces with time. Thermal consequences in the core under such conditions depend on the rate of reduction of coolant flow and the time of initiation of safety actions. Flow coast-down characteristics of the circuit depend on hydraulic inertia of the system. The inherent hydraulic inertia derived from the piping and pumping systems is so low that it can be negligible under practical conditions. Considerable inertial effects can be derived by mounting a flywheel on the pump shaft. This concept slows down the reduction of speed of the pump and thereby ensures the pumping function of the coolant in the circuit for a significant duration even after the trip of the pump motor. Coast-down duration of the pump is very important as a safety consideration, as it slows down the rate of the rise of temperatures in the core, which in turn allows comfortable time to initiate automatic safety actions. Larger coast-down time for pumps is desirable from the above considerations. However, since the coast-down time is directly linked to the size of the flywheel mounted on the pump shaft, long coast-down time leads to an expensive design. Apart from cost, accidental failure of a large-size flywheel can cause damage to nearby equipment and operating personnel. In this chapter a review of various considerations governing the selection of flow coast-down time for the pumping systems of a pool-type fast reactor is attempted. Some case studies highlighting these considerations with respect to a 500-MWe design of a typical fast reactor are also presented.

10.2 IMPACT OF COAST-DOWN TIME IN LOOP-TYPE SFR

The importance of flow coast-down time was understood in the early days of SFR design. A systematic study was carried out to investigate the impact of the selection of primary and secondary pump inertias on core cooling during a coast-down to a natural circulation event in a loop-type LMFBR

DOI: 10.1201/9781003283188-10

(Madni and Agrawal, 1980). Plant dynamic analyses were performed using the system dynamics code SSC-L (Agrawal et al., 1978) for the CRBRP design. It was observed from the studies that for any selected value of primary pump inertia (I_p), higher value of secondary pump inertia (I_s) is beneficial for better core flow and heat removal. Higher value of secondary inertia slows down flow decay in the secondary circuit. This causes more heat transfer to occur near the top of the intermediate heat exchanger (IHX), thereby shifting the thermal center upward. A higher location for the thermal center of IHX implies that there is more buoyancy head available for primary natural circulation. The thermal center of the core remains unchanged due to the power distribution being the same during normal operation as well as shutdown. However, for a given secondary system inertia, increasing of primary system inertia beyond a value causes a reduction in minimum core flow (Figure 10.1). This is due to the shifting of the thermal center of IHX downward due to the secondary flow getting heated within a short length.

For a given I_s, increasing I_p increases core flow up to a point, beyond which further increas reduces the minimum core flow. This is due to over-cooling, which destroys the buoyancy head. As an example, for $I_s = 1$, if I_p is increased beyond 3.2, this leads to core flow reversal, represented by the curve crossing zero of the ordinates. The reverse flow is of very small magnitude and is sustained. Inevitably, this is bad for core coolability. Eventually, the flow recovers and becomes positive again due to temperature rise in the core and reestablishment of buoyancy forces in the system. The duration of reverse flow increases with an increase in I_p. Flow reversal in the core leads to boiling if flow recovery is not fast enough. Figure 10.1 shows the corresponding map of maximum coolant temperature in the core.

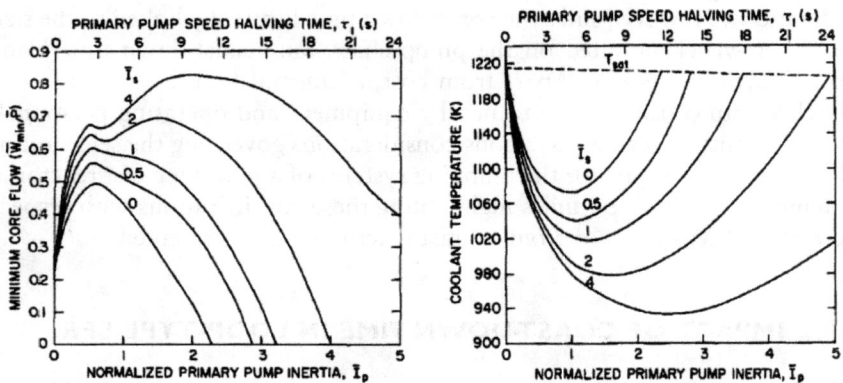

Figure 10.1 Minimum core flow and maximum hot channel temperatures—CRBRP.

Source: Madni, I.K., Agrawal, A.K., LMFBR System Analysis: Impact of Heat Transport System on Core Thermal Hydraulics. *Nucl. Eng. Des.* 62, 1980.

Under pipe rupture conditions, higher primary flow coast-down time results in faster draining of the system, leading to more severe consequences in the core. Nevertheless, in most of the loop-type reactor designs, the pipe rupture is regarded as a beyond-design-basis event by the provision of guard piping.

10.3 IMPACT OF COAST-DOWN TIME IN POOL-TYPE SFR

Figure 10.2 shows the map of core flow–to-power ratio (Q/P) for a typical pool-type fast reactor (Durham, 1977). Here, the abscissa is in terms of pump speed-halving times, i.e., the time taken by the pump to coast down from nominal speed to 50% of nominal speed. Since speed is proportional to flow, it is referred to as flow-halving time (FHT). In terms of halving times, the reverse flow effect appears to be more severe in pool-type designs than in loop-type designs. These results indicate the requirement of higher flow-halving time for a secondary sodium system compared to that for a primary sodium system to avoid flow reversal in the core.

Figure 10.3 presents the coast-down flow effects on initial cooling performance of pool-type Kalimer-600 reactor (Ji-Woong et al., 2012). The symbol, solid line, and dotted line represent the calculation results without additional coast-down flow, with coast-down times (CDT) 25 s and 50 s, respectively. The bottom graph represents Q/P as a function of time. Due to the imbalance between the decreased rate of flow and the core heat generation, the temperature fell off during the beginning stage without a peak of CDT = 25 s. For CDT = 50 s, the core experienced not only overcooling but

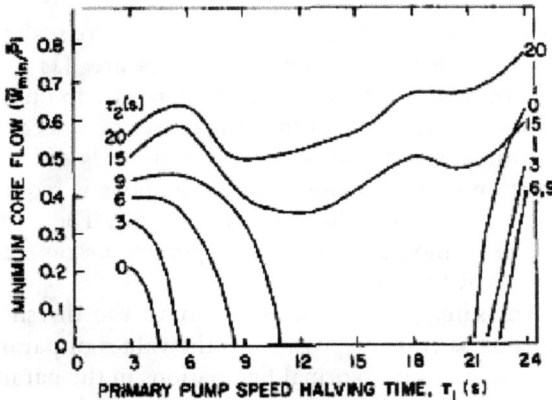

Figure 10.2 Minimum core flow in a pool-type reactor.

Source: M.E. Durham, Optimization of Reactor Design for Natural Circulation Decay Heat Removal in a Pool-Type LMFBR. In *Optimization of Sodium-Cooled Fast Reactors*. London: British Nuclear Energy Society, 1977.

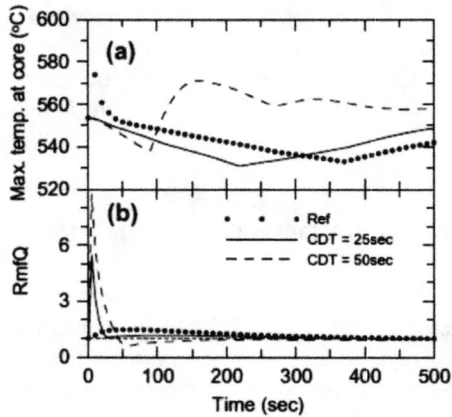

Figure 10.3 Effect of coast-down time on initial cooling performance.

Source: Han Ji-Woong, Jae-Hyuk Eoh, Seong-O Kim, Comparison of Various Design Parameters' Effects on the Early-Stage Cooling Performance in z Sodium-Cooled Fast Reactor. *Annals of Nuclear Energy* 40, 2012.

also undercooling due to the locally developed density-difference-driven flow. This resulted in negative and positive peaks of temperature for the initial duration for CDT = 50 s.

10.4 CONSIDERATIONS FOR DETERMINING FLOW COAST-DOWN TIME

There are several considerations for determining the coast-down of coolant in heat transport systems. The important factors are: (1) to ensure plant safety under loss-of-flow events; (2) to avoid reactor trip during short-term power fluctuations; and (3) to smooth the takeover of a decay heat removal system from a normal heat transport system under loss-of-heat-sink events. Careful investigation of each factor mentioned above is required to be carried out for suitably selecting the coast-down time. The first two considerations listed above impose conflicting requirements on fixing threshold values on reactor SCRAM parameters.

To ensure plant safety, it is desirable to have the threshold values on SCRAM parameters as close as possible to the values of parameters during normal operating conditions. Normal fluctuations in the parameters, uncertainties in measurement, and drift in setting of threshold values are considered when fixing the threshold values from the point of view of operating comfort. Under short-term power fluctuations, the pumping systems in the reactor would go through a transient phase leading to short-term flow reduction and pick-up. During this phase, in case any of the reactor's SCRAM thresholds are crossed, the reactor would get tripped automatically. Reactor trip under short-term power fluctuations can be avoided by (1) suitably fixing the threshold for SCRAM parameters such that SCRAM parameters do not cross the threshold value and (2) selecting longer flow coast-down time for pumping systems.

The third consideration listed above plays a key role in the loop-type design but not in the pool-type design. The loop-type design uses the concept of having the decay heat removal system in the secondary sodium circuit through water/air cooling of SG. Under loss-of-pumping-system conditions, the decay heat is envisaged to be removed through natural convection flow established in the primary and secondary sodium systems. The driving force for natural convection developed due to buoyancy depends on the thermal center difference between the heat source and the heat sink in the circuit. In the primary sodium circuit, the thermal center difference between reactor core and IHX is responsible for the buoyancy-induced driving force, whereas in the secondary sodium system the same is between IHX and decay heat exchanger or steam generator. During the coast-down phase of pumps, when the sodium flows in the primary and secondary sides are of comparable magnitude, the thermal center of IHX coincides with the geometrical center.

When the primary sodium flow is lower compared to the secondary flow, heat exchange to secondary takes place within a short distance from the primary inlet much above the geometric center of IHX, leading to an increased buoyancy head. This can be appreciated from the numerical studies carried out by for the FFTF IHX (Gunby, 1970). The temperature distribution across the length of the heat exchanger was assessed with various numerical schemes and compared with the exact solution for primary flow of 10% and secondary flow of 100%, as shown in Figure 10.4. It is evident that primary sodium loses its heat to secondary in the top 10% length from the hot end. Thus, the thermal center of IHX is much above the geometric center of IHX.

Figure 10.5 presents the case of secondary flow of 10% and primary flow of 100%. It is evident that the heat exchange is completed within 10% from the cold end, indicating that the thermal center is much below the geometric center in IHX. Therefore, depending on the relative magnitudes of primary and secondary sodium flows, the thermal center difference between reactor core and IHX can get altered. If it becomes negative, then the flow rate through the core may get extremely degraded and there may be flow reversal. Thus, the coast-down times of primary and secondary sodium systems should be such that conditions for core flow reversal are avoided.

From the above considerations, it is desirable to have higher flow-halving time for secondary compared to primary. In a pool-type design, the effect of thermal center difference between core and IHX gets masked to a great

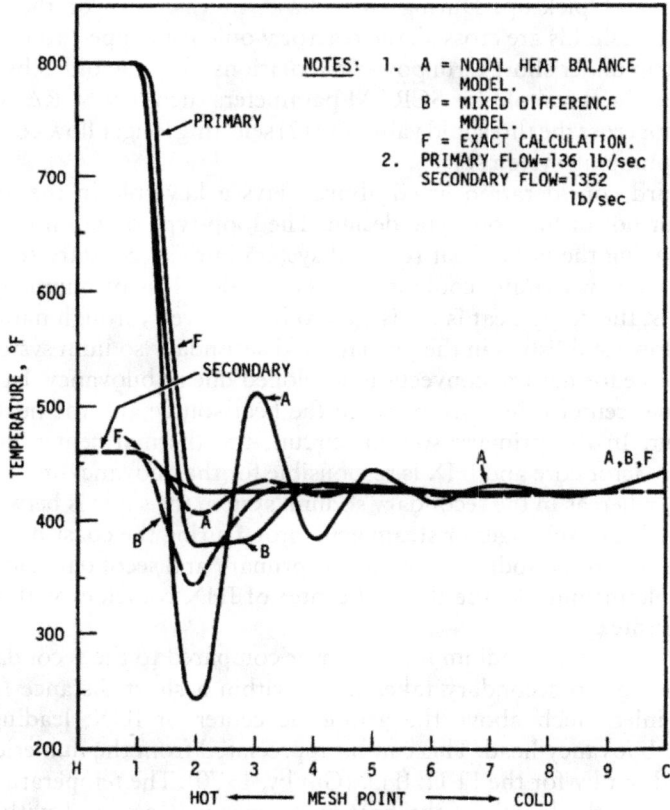

Figure 10.4 IHX Temperature distribution—10% primary flow FFTF.

Source: Gunby A.L., Intermediate heat exchanger modeling for FFTF simulation, BNWL-1367 UC-80, 1970.

extent by the thermal capacities of sodium pools, which delays the temperature evolution of the reactor inlet and the reactor outlet plenums, allowing for smooth establishment of stable natural convection flow in both primary as well as secondary. In the secondary sodium circuit, the natural convection flow will be established due to natural heat losses through pipelines. Following reactor SCRAM, reactor power reduces in the core and cold sodium starts coming out of fuel subassemblies. This causes cold shock on the structures above the core. To reduce this shock, one of the important automatic procedures adopted in the plant is to reduce speeds of sodium pumps after the reactor power reduces to decay power level. Lower coast-down time is beneficial to produce desirable results from this consideration. Nevertheless, under pump trip conditions, the hot shock on various components can be reduced by selecting a longer coast-down time.

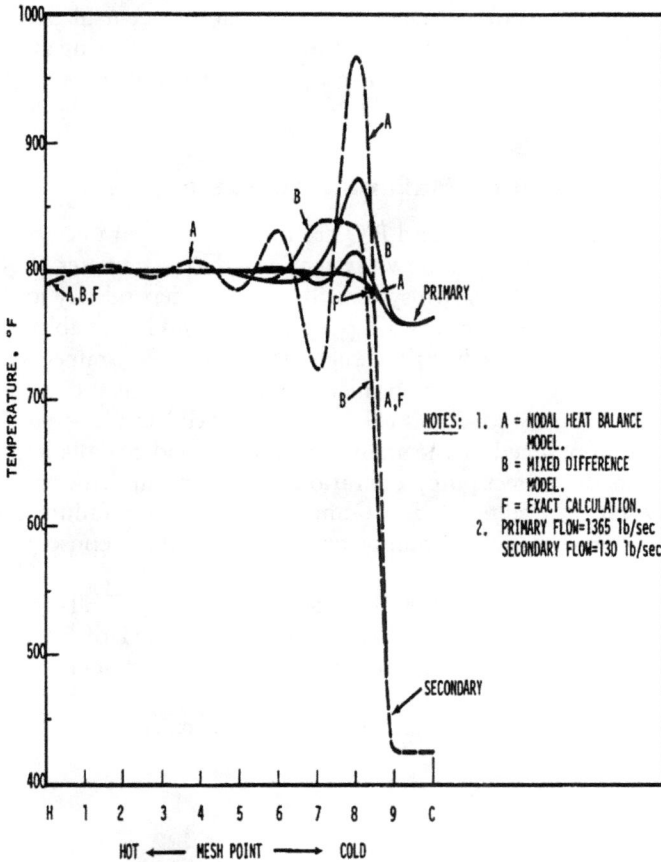

Figure 10.5 IHX Temperature distribution—10% secondary flow FFTF.

Source: Gunby A.L., Intermediate heat exchanger modeling for FFTF simulation, BNWL-1367 UC-80, 1970.

10.5 SCRAM THRESHOLD VERSUS COAST-DOWN TIME

Studies have been carried out for the PFBR reactor in India, generating a strong link between the SCRAM threshold and the coast-down time. Thermal hydraulic analysis of a power failure event has been carried out using DYANA-P code (Natesan et al., 2015). This event is simulated by assuming instantaneous dry-out of all SG and setting the drive torques of primary and secondary sodium pumps to zero. Thus, the primary and secondary pumps coast-down is governed by their inertia. Among the list of SCRAM parameters considered in the PFBR design, those that evolve to trip the reactor during a power failure event are low primary pump speed (N_P), high power-to-flow ratio (P/Q), high central SA sodium outlet temperature

(Θ_{CSASM}), and high coolant temperature rise across central SA ($\Delta\Theta_{CSAM}$). Θ_{CSAM} and $\Delta\Theta_{CSAM}$ evolve at the same time for events occurring at full-power conditions. Therefore, for practical purpose of analysis, these two parameters can be treated as one.

10.5.1 FHT Effect on Maximum Temperatures

The SCRAM parameters N_P and P/Q are the first two parameters appearing and are connected to first shutdown system. The third parameter, Θ_{CSAM} is connected to the second shutdown system. To ensure independency in the design of the two shutdown systems, Θ_{CSAM} parameter should be capable of ensuring plant safety without exceeding the design safety limits. A parametric study has been carried out by varying the primary coast-down time and event analysis has been carried out. Secondary FHT is considered constant at 4 s in this study.

Primary pump speed is the main parameter used for plant protection against events that affect pump operation, namely pump trip, power failure, and pump seizure events. Out of pump trip and power failure events, the later event (during which all pumps trip) includes all the consequences of a single pump trip event.

Variation of minimum core flow against primary FHT is shown in Figure 10.6. It can be observed that initially up to a FHT of 5 s, an increase in FHT causes reduction in minimum core flow reached during the

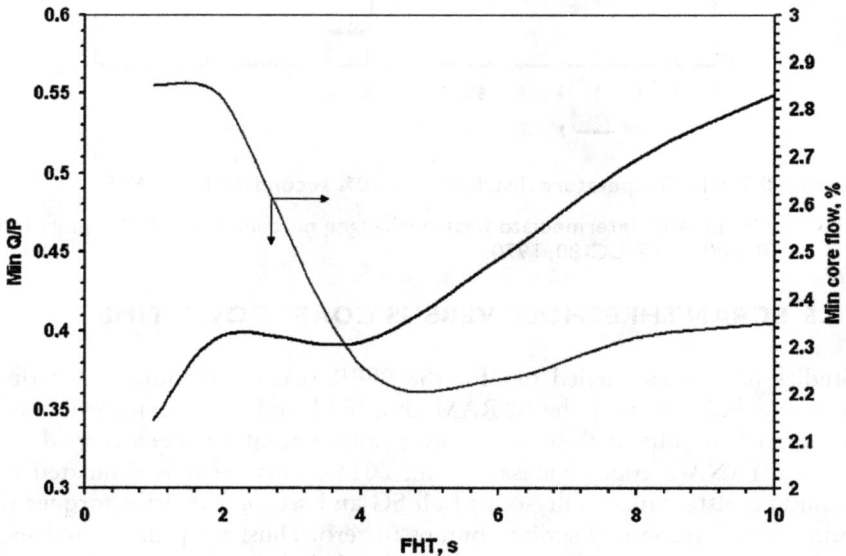

Figure 10.6 Minimum Q/P and core flow as function of primary FHT.

Source: Natesan K. et al., Significance of Coast Down Time on Safety and Availability of a Pool Type Fast Breeder Reactor. *Nuclear Engineering and Design* 286, 2015.

transient. This is due to the interaction of buoyancy developed in the primary sodium circuit during the transient. However, this reduction does not affect the Q/P ratio significantly. This is because the time at which minimum core flow is reached in the core gets extended with an increase in FHT. Therefore, power produced in the core at the instant of minimum core flow reduces with an increase in FHT. However, from the point of view of having increased core flow during the transient with increase in FHT, it is desirable to have it more than for 5 s. To keep the clad hotspot temperature below the limit of 800°C, the FHT required in the primary sodium circuit is 8 s. Thus, from the point of view of core safety, primary FHT of 8 s is essential.

10.5.2 FHT to Avoid SCRAM for Short Power Failure

When power failure occurs in the plant, all pump motors get tripped, and flows in the respective circuits start reducing. This in turn causes the rise in coolant temperature. If power comes back within a short time, the coolant flows, and hence the coolant temperatures, will be restored. If any of the SCRAM parameters crosses its trip threshold during this transient, then reactor SCRAM happens, and restoration of power cannot ensure continued plant operation. The reactor will have to be taken to a cold shutdown state and plant start-up will have to be initiated for bringing the reactor back to operation. This significantly affects the plant availability. Therefore, it is desirable to avoid reactor SCRAM during short-term power disturbances lasting for a few seconds. This can be achieved if the reactor SCRAM parameters do not increase to their respective SCRAM threshold values during the short-term power failure conditions. To accomplish this, coast-down times of coolant circuits must be chosen accordingly. P/Q is a parameter that gives protection to a plant under loss of flow and overpower events. Θ_{CSAM} is the one that has a major impact on the hot spot sodium and clad temperatures. The Np parameter threshold can then be fixed accordingly.

Studies showed that by raising the Θ_{CSAM} temperature threshold from 10°C to 15°C, the minimum FHT for a power failure duration of 2 s went from 24 s to 33s. If the duration of power failure is taken as 0.5 s, then minimum FHT comes down from 8 s to 5 s.

10.6 SECONDARY PUMP FHT

In pool-type reactors with a decay heat removal system in the pool, there is no concern for the natural convection flow setting up in secondary. Hence, it must be decided based on the thermal shocks because of a secondary pump trip and/or feedwater pump trip event. In loop-type reactors, which do not have a decay heat removal system in the hot pool, the impact of setting up natural convection in the primary needs a higher flow in the secondary, and hence the FHT in the secondary must be more than in the primary.

10.7 PRIMARY FHT FOR UNPROTECTED LOSS OF FLOW

SFRs have a double-walled reactor vessel to avoid the loss of coolant in case of a leak from the primary sodium circuit. Also, in loop-type reactors, the pipes connecting the reactor to the IHX and the pump are double-walled. In view of this, chances of a loss-of-coolant event from primary are very low. Hence, loss of flow (LOF) becomes a more important event in terms of safety considerations. Current safety regulations demand redundancy and diversity in both the shutdown and decay heat removal systems to assure high safety reliability. Hence, we have two shutdown systems based on diverse concepts. However, older reactors had only one shutdown system and one decay heat removal path constructed. The saving grace was that they were low-power reactors in which all the feedback reactivity coefficients were negative.

The best example is the RAPSODIE reactor, where a test was conducted at 50% power by cutting off power supplies to primary sodium pumps, secondary sodium pumps, and air blowers to the terminal sodium to air exchangers and not allowing the control rods to drop (unprotected LOF). The reactor stabilized at a low power of ~10kW, and the test was terminated after 15 mins (Figure 10.7). The large flow-halving time (17s at full power)

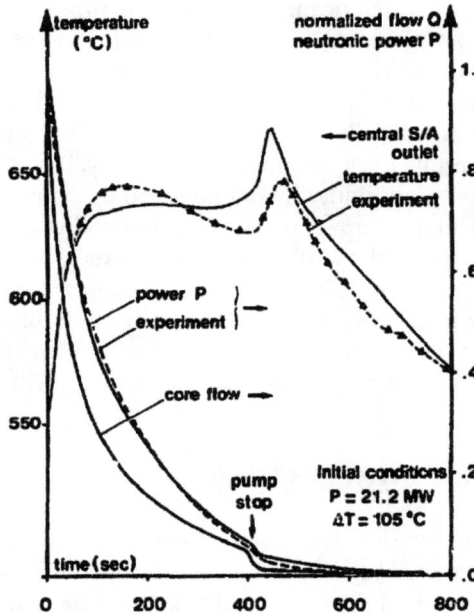

Figure 10.7 RAPSODIE ULOF test results.

Source: Liquid metal cooled reactors: experience in design and operation, IAEA, VIENNA, 2007, IAEA-TECDOC-1569.

of the pump has been found to play an important role in reactivity dynamics, by bringing in sufficient negative feedback reactivity due to structural expansions in reactor assembly in this case.

An analysis was carried out for the FBTR, which is based on RAPSODIE design but with a steam generator instead of a terminal air exchanger (Vaidyanathan, 2010). It was observed that the reactor stabilized at low power at a temperature higher than in RAPSODIE. Variations in reactor power, hot spot clad, and fuel temperatures, besides the contributions of different feedback reactivities, are presented in Figure 10.8. The grid plate expansion is initially positive due to the initial fall in reactor inlet temperature. This fall is governed by the relative flow rates of primary and secondary sodium in IHX. Later, this feedback becomes negative due to a rise in temperature and contributes significant negative feedback. The other

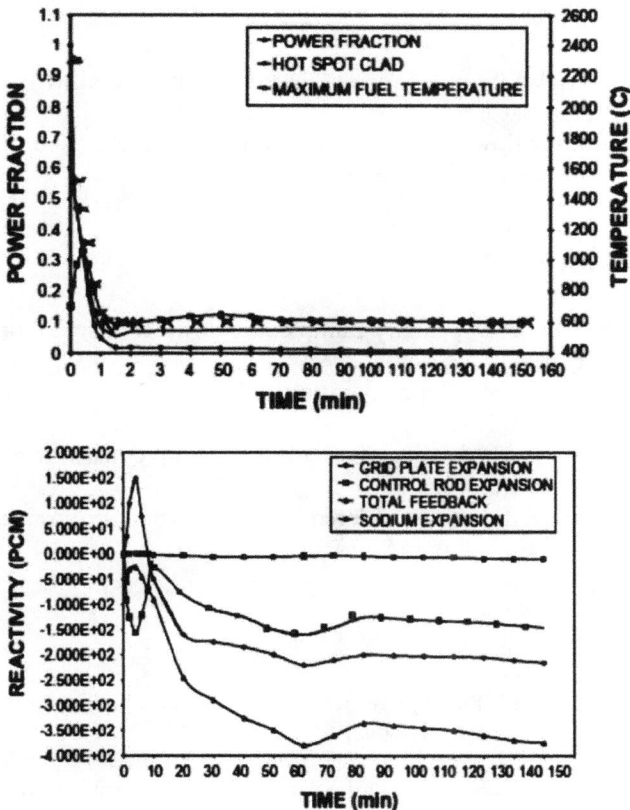

Figure 10.8 Reactor parameter evolution for FBTR in ULOF.

Source: Vaidyanathan G. et al., Dynamic Model of Fast Breeder Test Reactor. *Annals of Nuclear Energy* 37, 2010.

Table 10.1 Primary FHT in Different Reactors

S. No.	Reactor/Type/Country	No of Shutdown Systems	FHT (s)	Reference
1	RAPSODIE/Loop/ France	1	17	Essig, 1985
2	FBTR/Loop/India	1	17	Vaidyanathan, 2010
3	MONJU/Loop/Japan	1	6	Yamada and Kitamura, 2004
4	PFR/Pool/UK	1	10	Jensen and Olgaard, 1995
5	PHENIX/Pool/France	1	8	Natesan et al., 2015
6	SUPER PHENIX/ Pool/France	2	50	Gouriou et al., 1982
7	Kalimer-600/Pool/ South Korea	2	10s	Chang et al., 2011
8	CEFR/pool/China	2	10 s	Vasile, 2017

contributor is the sodium expansion because of the overall rise in the average primary temperatures.

Primary sodium flow-halving time in different reactors compiled from literature is presented in Table 10.1.

ASSIGNMENT

1. What is the impact of primary pump coast-down time in a loop-type and pool-type SFR?
2. What are the important considerations while deciding the primary pump coast-down time?
3. Describe the relationship between primary pump coast-down and safety set points for a reactor trip.
4. What aspects need to be considered in fixing secondary flow coast-down time in loop- and pool-type SFRs?

REFERENCES

Agrawal A.K., et al. (1978), An Advanced Thermo Hydraulic Simulation Code for Transients in LMFBRs (SSC-L Code) BNL-NUREG-50773.

Chang W.P., Kwon Y.M., Jeong H.Y., Suk S.D. and Lee Y.B. (2011), "Inherent safety analysis of the KALIMER under a LOFA with a reduced primary pump halving time", *Nucl. Eng. and Technol.*, Vol. 43, No. 1; https://doi.org/10.5516/NET.2011.43.1.063

Durham M.E. (1977), Optimization of reactor design for natural circulation decay heat removal in a pool-type LMFBR, Optimization of Sodium-Cooled Fast Reactors, British Nucl. Energy Soc., London.

Essig C. (1985), Dynamic behaviour of RAPSODIE in exceptional transient experiments. In: Proceedings ANS International Topical Meeting on Reactor Safety, Knoxville, USA.

Gouriou A., et al. (1982), Dynamic behavior of super PHENIX reactor under unprotected transient. In: Proceedings LMFBR Safety Topical Meeting, Lyon.

Gunby A.L. (1970). Intermediate heat exchanger modeling for FFTF simulation, BNWL-1367 UC-80, Reactor Technology.

Jensen S.E. and Olgaard P.L. (1995), Description of prototype fast reactor at Dounreay. In: NKS/RAK-2(95) TR-C1. Rio National Laboratory, Roskilde, Denmark.

Ji-Woong H., Eoh J. H. and Kim S. O. (2012), Comparison of various design parameters' effects on the early-stage cooling performance in a sodium-cooled fast reactor, *Ann. Nucl. Energy* 40, pp. 65–71. doi: 10.1016/j.anucene.2011.10.006

Madni I.K., Agrawal A.K. (1980), LMFBR system analysis: impact of heat transport system on core thermal hydraulics. *Nucl. Eng. Des*, Vol. 62, pp. 199–218

Natesan K., Velusamy K., Selvaraj P. and Chellapandi P. (2015), Significance of coast down time on safety and availability of a pool type fast breeder reactor, *Nucl. Eng. Des*, Vol. 286, pp. 77–88; http://dx.doi.org/10.1016/j.nucengdes.2015.01.021

Vasile et al. (2017), Recent Activities of the Safety and Operation Project of the Sodium-Cooled, Fast Reactor in the Generation IV International Forum, IAEA-CN245-133, *International Conference on Fast Reactors and Related Fuel Cycles: Next Generation Nuclear Systems for Sustainable Development FR17*, Yekaterinburg.

Yamada F. and Kitamura K. (2004), "*Realistic safety margin analysis of 'MONJU' based on plant performance measurements*," 12th International Conference on Nuclear Engineering, Arlington, Virginia, USA, April 2004.

Chapter 11

Plant Protection System

11.1 INTRODUCTION

Reactor protection against various design basis events that occur in the plant is ensured by tripping the reactor based on the values of a certain selected set of plant parameters, known as SCRAM parameters, crossing their specified thresholds. An adequate number of SCRAM parameters are required in the plant protection system to limit the consequences of various design basis events such that the safety criteria are respected. Meanwhile, the number of such parameters should also be kept at a minimum to increase plant availability. The plant protection system should have reliable instruments to generate signals for a timely shutdown of the plant based on evolution of SCRAM parameters. To design the instrumentation and other systems of the plant protection system, the maximum permissible delay for actuation of these systems needs to be determined. The permissible delay time should be such that the consequences following various design basis events are limited within the specified safety criteria. One-dimensional system simulation codes, DYNAM and DYANA-P, developed for FBTR and PFBR, respectively, have been employed for this purpose. Using these codes, various design basis events envisaged in the plant have been investigated to obtain the flow and temperature evolutions in all parts of the plant. The results of these studies have been analyzed and reactor trip action proposed on their basis. The maximum permissible delays for the actuation of a reactor trip by various SCRAM parameters have been worked out to ensure plant safety under all transient conditions. This chapter presents the results of the thermal hydraulic analyses that led to the design of the plant protection system for a loop-type and pool-type SFR.

11.2 LIMITING SAFETY SYSTEM SETTINGS (LSSS) FOR FBTR

FBTR, as explained earlier, is a loop-type experimental reactor, with both primary loops in parallel, each having an independent secondary loop with

DOI: 10.1201/9781003283188-11

two SG modules (Vaidyanathan et al., 1987). It has a shutdown system comprising six control rods. The procedure of arriving at the safety set points of different parameters is described in the following.

11.2.1 Safety Signals and Settings

Neutron flux–related measurements such as neutronic power, reactivity, period, flow-related power/flow ratio, core temperatures, and delayed neutron activity form the main instrumentation to provide safety trip signals. Neutron flux measurement by fission chambers and compensated ion chambers is processed through analog circuitry to get log count rate, reactor period (Tp), linear power (Lin P), log power (Log P), and reactivity (ρ). The temperatures measured at the outlet of fuel subassemblies (Θi) are processed by a computer to arrive at a global mean outlet temperature (Θm) and a global mean temperature gradient ($\Delta\Theta m$). The reactor flow signal (Q) is generated by summing up the signals of flow in each loop. This, along with Lin P, is used to generate a P/Q signal. Delayed neutron detector (DND) signal is used to detect delayed neutrons in sodium due to fuel failure.

All electronically processed analog signals are susceptible to variations due to drift, temperature, power supply fluctuations, calibration accuracy, nonlinearity, etc. Over and above this, adequate control margin must be kept around the rated value for operator control. Such considerations would dictate a set point away from the rated value. Two approaches are available to decide the set point for any parameter. In the first approach, it is assumed that all the variations of the signal occur in the same direction, without being compensated by one another and control margin added to arrive at the trip setting inaccuracy. This set point would then ensure that there are no false trips of the reactor, which can affect plant availability.

In the second approach, the variations in the signals due to various considerations are computed based on a statistical approach with a confidence level of 99% (2 sigma). To this the control margin is added, and this is the trip setting inaccuracy. The set point is fixed at twice the trip setting inaccuracy to minimize false trips. For example, temperature measurement error for FBTR was found to be +5°C, and the set point for the trip was fixed at 10°C. The set point obtained in this approach would give greater plant availability. The set point arrived at is analyzed for the safety margin it provides for different events.

11.2.2 Limiting Safety System Settings (LSSS) Adequacy

The DYNAM code was used to evaluate the adequacy of the set points arrived at based on the statistical approach discussed above. The limits to be observed are the maximum fuel temperature of 2566°C and hot spot clad temperature of 800°C. The incidents that cause reactor power and/or temperature increase are failure/seizure of one primary pump, total power failure, and inadvertent withdrawal of one control rod. For incidents on secondary sodium and steam generator, an increase in reactor inlet temperature results in a decrease of reactor power (due to overall negative feedback reactivity coefficients) and hence the temperature drop at the core outlet. In evaluating the adequacy of the set points, it was also examined whether there are a minimum of two diverse signals to ensure a reactor trip for the different events. Looking at diversity, it may be noted that reactor period (Tp), log power (Log P), and reactivity (ρ) are derived from neutron flux (Lin P), and hence a combination from this set cannot be called diverse.

Table 11.1 gives the safety parameters, their set point threshold from maximum availability considerations, the time at which safety action takes place, and values of the hot spot clad and fuel centerline temperatures at the time of trip, for the different events. It may be noted that for each parameter, two sets of results are generated if a trip occurs at the nominal threshold or if it occurs at more than the nominal threshold due to measurement errors. For example, for sodium outlet temperature of a fuel subassembly, the nominal threshold is 10°C, and with a measurement error of + 5°C, a reactor trip

Table 11.1 Evaluation of LSSS

Event	Parameter	Threshold	Time to reach threshold, S	Hot spot clad temperature, °C	Max. Fuel Temperature, °C
Inadvertent withdrawal of one control rod	ρ	± 10 pcm	2.1	648.5	2321
			2.2	650.0	2330
	Θ m	10°C	7.0	668.0	2470
			9.5	677.0	2540
	Lin P	1.145	8.5	673.6	2508
			11.0	685.0	2600
	P/Q	1.3	8.5	673.6	2508
			11.5	685.0	2600
	Log P	1.85	>20.0	>730.0	>3000
			>20.0	>730.0	>3000
Failure of one primary pump	P/Q	1.3	8.0	690.0	2100
			Threshold not reached	-----	-----
	Θ m	10°C	6.5	680.0	2180
			9.0	690.0	2100
Total power failure	Θ m	10°C	8.5	712.0	2096
			11.0	738.0	2021
	P/Q	1.3	7.0	700	2153
			Threshold not reached	------	-------

could happen when the temperature has risen by 15°C. For the event of inadvertent withdrawal of one control rod, reactivity and temperature signals can ensure safety. It may be noted that the parameters Lin P, Log P, and reactivity are derived from the same neutronic signal and cannot be treated as diverse.

The reactivity parameter also has a negative threshold. The negative threshold is to take care of heavy blockage in the fuel assembly that would result in a temperature increase inside the assembly but due to low flow (resulting in higher transportation time) may not be felt at the outlet to be measured by the core outlet thermocouples. For low-power SFRs, the reactivity feedback coefficients are negative throughout the core, and the temperature rise within would reflect as a negative reactivity that can be used to trip the reactor. This is essentially based on the fuel melting incident in the Enrico Fermi fast breeder reactor, where an operator raised the control rod to compensate the negative reactivity due to boiling sodium (not indicated by temperature measurements) inside the core (APDA-233, 1966). The boiling was apparently caused by the obstruction to a fuel assembly flow by a plate, which got detached and was carried to the core inlet. Based on this event, all subsequent reactor designs ensure multiple radial inlets besides the axial one to minimize the probability of flow blockage.

For the event of one primary pump trip and total power failure, we have only the temperature signal to give the trip. This is because the P/Q signal (power/reactor flow) reaches a peak below the threshold set point and then falls due to faster reduction of power than flow. This is essentially due to a power drop due to negative feedback reactivity caused by flow reduction resulting in a temperature increase. One could reduce the set point of P/Q, but then it could result in false trips and affect plant availability. Thus, in place of P/Q, the signal Q_{mini} was proposed. If the primary sodium flow in any of the loops goes below the set point, the signal trips the reactor. While evaluating this it was found that this provided safety at higher powers, but at lower powers, due to hydraulic coupling between the two primary loops for the events of one primary pump failure/seizure, the operating pump was able maintain the flow significantly above the set point. To overcome this, a differential flow of the primary flow in the two loops was proposed and was found to ensure safety. In regards to the set point on the period, the evaluation is based on inadvertent withdrawal of one control rod at low power.

11.3 LIMITING SAFETY SYSTEM SETTINGS FOR PFBR

PFBR is a pool-type reactor with two primary loops and two secondary loops, and each secondary loop has four modules of SG. After FBTR, the regulatory authorities drew up a detailed safety criteria for sodium-cooled fast reactors. Safety criteria relevant to plant protection system as mentioned in the AERB document (AERB, 1990) are listed below.

11.3.1 Design Basis Events

Events that affect the performance of the plant are broadly classified as design basis events (DBE) and beyond design basis events (BDBE). The DBE have the frequency of occurrence more than 10^{-6} per reactor-year (ry) (IAEA, 1985). The DBE are further classified into four categories according to the frequency of their occurrence, namely Category 2 ($>10^{-2}$/ry), Category 3 ($\leq 10^{-2}$/ry but $>10^{-4}$/ry), and Category 4 events ($\leq 10^{-4}$/ry but $>10^{-6}$/ry), with Category 1 being all the planned operations. The BDBE have the frequency of occurrence less than or equal to 10^{-6}/ry.

Various events and their categorization have been identified for the plant based on literature available on various fast reactors and probabilistic safety analysis. Among the various concerns addressed in the analysis of DBE, the major aspects are fuel pin integrity and individual component structural integrity. Thermal hydraulic analyses have been carried out only for a selected set of DBEs called the enveloping events. The thermal hydraulic consequences following these events envelop the consequences following all the other events considered in the design. The plant protection system has been designed based on the analysis of the enveloping events such that the safety criteria are respected, and the system integrity is ensured for the entire lifetime of the plant.

11.3.2 Core Design Safety Limits

- The reactor core and associated coolant, control, and protection systems shall be designed with appropriate margins to ensure that the specified design safety limits (DSL) such as maximum temperatures of fuel, clad, and coolant are not exceeded during all operational states (including anticipated operational transients). DSL for various categories of events have been arrived at for different components such that the resulting cumulative structural damage estimated for the full life time of the plant is well below the permissible limits. Cumulative damage is estimated by considering a credible number of occurrences of different events of various categories. DSLs have been prescribed for cold pool, hot pool, average coolant hot spot, fuel clad hot spot, storage subassembly clad, and fuel hot spot temperatures for the four categories of events (Natesan et al., 2012).
- The various coolant systems and associated control, protection, auxiliary, and sodium heating systems shall be designed to have sufficient capacity with adequate margins and redundancy to remove the heat from the core and transport it to the ultimate heat sink, without exceeding the specified limits, under all operational states of the reactor and postulated accident conditions.
- For all DBEs, the first SCRAM parameter should limit the consequences within the specified category DSL of the event, and the second

SCRAM parameter should limit the consequences within the next, higher-category DSL. However, for the category 4 DBE, two SCRAM parameters should limit the consequences within the category 4 DSL.

11.3.3 Selection of SCRAM Parameters

The most important plant measurements are neutron flux (Φ), sodium temperatures at the core inlet (Θ_{RI}) (measured at the PSP suction), central SA outlet (Θ_{CSAM}), other individual fuel SA outlets (Θ_{I}), core flows (QPP) (measured at the PSP discharge), and the delayed neutron detector (DND) flux. Some of the important signals derived from these measurements are reactor power (Lin P), period (τ_{N}), reactivity (ρ), power-to-flow ratio (P/Q), group mean of SA sodium outlet temperature (Θm), deviation of individual SA sodium outlet temperature from an expected value ($\delta\Theta i$), the mean core temperature rise ($\Delta\Theta\ m$), and temperature rise in central SA ($\Delta\Theta_{CSAM}$). The primary pumps are operated at 100% flows at all powers. Hence, a parameter of primary pump speed (N_P) is also added to the trip signals.

Overpower events can be detected by the power, reactivity, or reactor period and temperature rise in the core. Undercooling events can be detected by power-to-flow ratio, outlet temperature rise in the subassemblies, and reactivity. DND serves for detecting fuel clad failures.

The transient evolutions of the coolant, clad and fuel temperatures are affected by the delays encountered by the measuring instruments, signal processing, logic circuits, SCRAM release electromagnet, and the rod drop time, over and above the process delay reaching a trip threshold. Once the process value of a SCRAM parameter crosses LSSS, the reactor SCRAM by the actual movement of control rods will be started after a specific time delay. This time delay is the total sum of response time of instrumentation, delay associated with the trip logic circuit, and electromagnetic clutch release time. The plant protection system should be designed such that the total delay time associated with various subsystems is less than that permissible time, to limit the consequences within DSL.

11.4 SHUTDOWN SYSTEM

The automatic shutdown of PFBR is accomplished by the dropping of control rods in two independent and diverse shutdown systems, namely SDS-1 and SDS-2, unlike FBTR. SDS-1 consists of nine control and safety rods (CSR), and SDS-2 consists of three diverse safety rods (DSR). CSR are used for shutdown and power control whereas, and DSR are used only for shutdown function. Control logic ensures that only one CSR or DSR can be selected at a time, and the CSR are raised only after all the DSR are fully raised. These two shutdown systems are connected to two independent safety logics and different cable routings, to reduce the probability of common-cause failures

(Figure 11.1). Diversity has been ensured in the design of sensors, safety logic, drive mechanisms, and absorber rods (Senthil et al., 2005). SCRAM release electromagnet operates in argon space and sodium environments in SDS-1 and SDS-2, respectively. Apart from this, oil dash pot and sodium dash pot designs are adopted in SDS-1 and SDS-2. A diverse set of SCRAM parameters trigger the dropping of rods in SDS-1 and SDS-2. Among the safety logic parameters, τ_N, Lin P, ρ, P/Q, Θ_{RI}, and N_p are connected to SDS-1. The parameters Θ_{CSAM}, $\Delta\Theta_m$, $\Delta\Theta_{CSAM}$, and $\delta\Theta i$ are connected to SDS-2. DND signal is connected to both SDS due to the unavailability of another parameter for the detection of random failures of fuel pins.

The drive mechanism of CSR (CSRDM) is designed for a total drop time of 1 s. This includes 0.1 s delay for electromagnetic clutch release and an

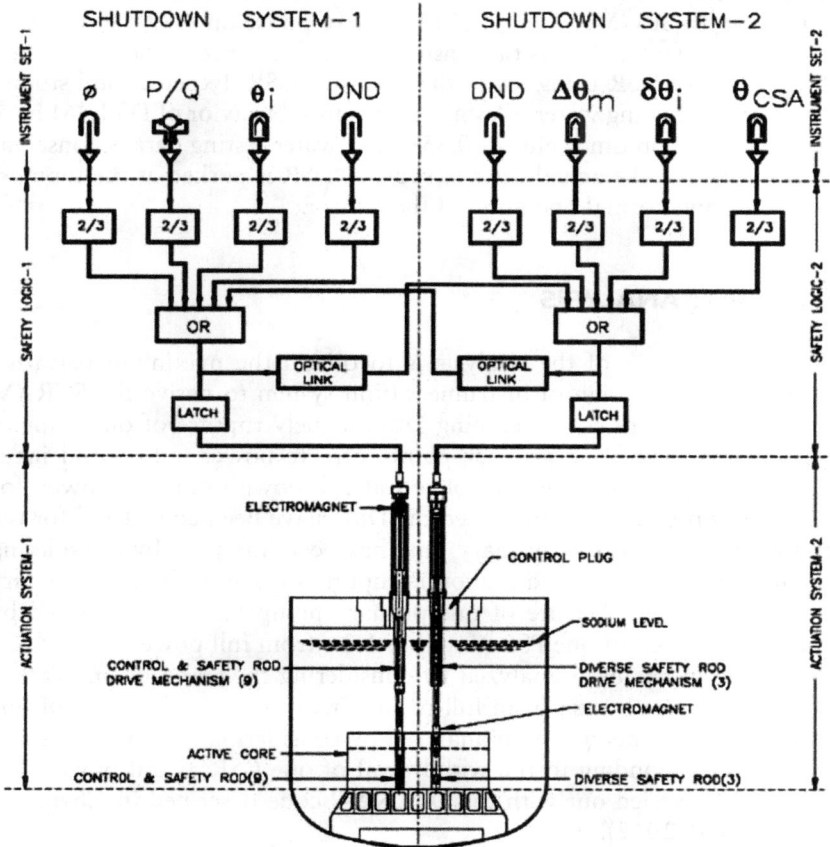

Figure 11.1 Shutdown system logic for PFBR.

Source: Senthil Kumar C, A. John Arul, Om Pal Singh, K. Suryaprakasa Rao, Reliability analysis of shutdown system. *Annals of Nuclear Energy* 32, 2005.

additional delay of 0.2 s due to safe shutdown earthquake (SSE). Thus, the time required for its maximum free fall to dash pot from its topmost position is 0.7 s. It may be highlighted that the drop time of CSRDM observed in the experimental studies in stagnant sodium is 470 ms. This value corresponds to the worst-condition data obtained from a series of experiments with aligned and misaligned orientations of mechanism and subassembly. Further, deceleration of rods in the dash pot takes place with a braking time of 220 ms. The position of CSR in the core during normal operating conditions depends on the burnup composition of the core. For conservative analysis, the position considered is that corresponding to the end of equilibrium cycle (EOEC) such that the rod movement is maximum during SCRAM.

In the case of DSR, the electromagnetic clutch release delay is 0.1 s and deceleration effect due to SSE of 0.1 s. CSRs are actuated by safety logic with a fine impulse test (SLFIT) whereas DSRs are actuated by pulse-coded safety logic (PCSL) (Misra et al., 2014). The response times of the two logic systems are different. This is the reason for the difference in the delay times for actuation of DSR being lower than that for CSR. Experimental studies carried out in flowing water to establish the drop behavior of DSRDM have given a worst drop time value of 0.85 s. The water testing data is conservative compared to the actual drop in sodium. DSR is parked at the topmost location during normal operation of the reactor.

11.5 EVENT ANALYSIS

The main objective of the analysis is to obtain the maximum permissible delay for the design of instrumentation system to derive the SCRAM parameters. A set of six enveloping DBE, namely rupture of one primary pipe, seizure/trip of one primary pump, off-site power failure and inadvertent withdrawal of one control rod at full power and low power, for which the consequences can exceed the DSL, have been considered for the analysis. Rupture of one primary pipe has been analyzed by considering instantaneous double-ended guillotine rupture of one of the four primary pipes at full power. Seizure of one primary pump has been analyzed by reducing the speed of one PSP to null in 1.0 s from full power. Trip of one primary pump has been analyzed by considering the speed of one PSP to coast down by its inertia from full power. Inadvertent withdrawal of one control rod incidents has been analyzed by considering external reactivity addition corresponding to the withdrawal of one CSR at full power. The analysis was carried out with the DYANA-P code described in Chapter 9 (Natesan et al. 2012).

As indicated earlier, all electronically processed analog signals are susceptible to variations due to drift, temperature variation, supply fluctuations, accuracy of calibration, nonlinearity, and so on. Similarly, set point values are also susceptible to variation. The net impact of these effects is the shifting of the LSSS of the SCRAM parameters from their nominal set points. The maximum deviations of the set points from their nominal settings due to all the above-mentioned effects have been estimated. Control margins and total error are computed based on statistical variations.

Drift in the measurement channel due to maximum possible variation in the temperature of electronic cabinets is also considered in this estimation. Thus, in the present analyses, reactor trip is triggered by the SCRAM parameters when the measured value of parameter exceeds the sum of threshold value and its deviation.

The reactor protection system should be able to safely shut down the reactor before the consequences following the events exceed their respective category limits. All the events have been analyzed without considering any safety action to determine the time at which the process values of various SCRAM parameters reach their respective thresholds. Analysis has also been carried out by considering reactor SCRAM by insertion of rods by SDS-1 or SDS-2 at various instants following the events to estimate the maximum allowable SCRAM initiation time such that the DSL are not exceeded. While analyzing SCRAM by SDS-1, single failure of one CSR is considered, and the shutdown is effected by dropping eight CSRs. Similarly, while analyzing SCRAM by SDS-2, a single failure of one DSR is considered, and the shutdown is effected by dropping two DSRs.

Flow in the P/Q parameter corresponds to the flow in the small by-pass passage provided for the mounting of eddy current (EC) flow meters. Transient evolution of flow in this passage has been calculated by including hydraulic inertia of this passage.

In PFBR, routine start-up of the reactor is envisaged with primary and secondary sodium pumps running at full speeds, corresponding to the targeted power level throughout the power-raising procedure. PFBR operation adopts a philosophy of coast-down of speeds of primary and secondary sodium pumps to 20% nominal following reactor SCRAM. This is to reduce cold thermal shock on components. Feedwater flow through SG is manipulated to control secondary sodium temperature at the outlet SG. This control action reduces feedwater flow through SG to 15% (minimum value) automatically following SCRAM. These scenarios are considered in the present analysis. Analysis has been carried out for all six events indicated above with SCRAM by SDS-1 and SDS-2. As an example, for the case of off-site power failure where the trip parameter is N_p, the evolution of clad hot spot temperatures and arriving at maximum permissible time for instrumentation and control (I&C) is shown Figure 11.2.

Tables 11.2 and 11.3 present the overall results of the analysis to arrive at an estimation of maximum permissible delay time for instrumentation and control (I&C) to limit the consequences below DSL for parameters linked to SDS-1 and SDS-2, respectively.

The instrumentation and control system design considers the maximum delay time available as presented in Tables 11.2 and 11.3, and the systems

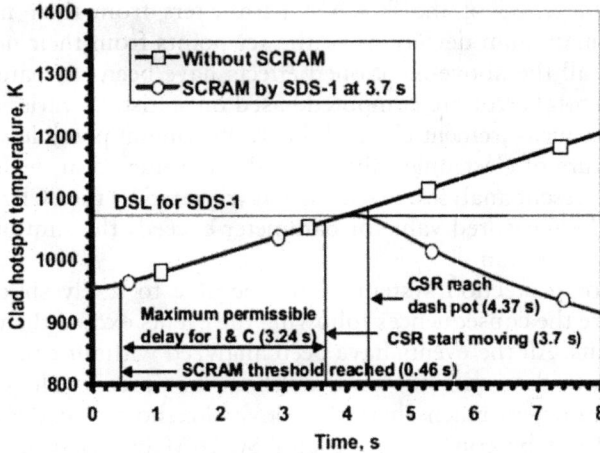

Figure 11.2 Permissible delay for N_P parameter for off-site power failure.

Source: Natesan K., et al., Plant Dynamics Studies Towards Design of Plant Protection System for PFBR. *Nuclear Engineering and Design* 250, 2012.

Table 11.2 Maximum Permissible Delay Time for SDS-1 Parameters

S No.	Event (category)	Trip parameters (DSL)	Permissible delay time for I&C, s
1	One control rod withdrawal at full power (2)	Lin P (FHST < 2,926 K)	6.10
2	One control rod withdrawal at low power (2)	Lin P (FHST < 2,926 K)	6.69
3	Trip of one PSP (2)	N_P (CHST < 1,073 K)	6.44
4	Seizure of one PSP (3)	N_P (CHST < 1,173 K)	0.44
5	Off-site power failure (2)	N_P (CHST < 1,073 K)	3.24
6	Primary pipe rupture (4)	P/Q (AVSHS < 1,213 K)	2.45

Table 11.3 Maximum permissible delay time for SDS-2 parameters

S No.	Event (category)	Trip parameters (DSL)	Time available for I&C, s
1	One control rod withdrawal at full power (2)	Θ_{CSAM} (FHST < 2,926 K)	6.17
2	One control rod withdrawal at low power (2)	$\Delta\Theta_{CSAM}$ (FHST < 2,926 K)	5.62
3	Trip of one PSP (2)	Θ_{CSAM} (CHST < 1,173 K)	14.33
4	Seizure of one PSP (3)	Θ_{CSAM} (AVSHS < 1,213 K)	DSL not crossed
5	Off-site power failure (2)	Θ_{CSAM} (CHST < 1,173 K)	5.39
6	Primary pipe rupture (4)	Θ_{CSAM} (AVSHS < 1,213 K)	1.45

have been found to meet this requirement. This ensures that safety of the reactor is assured for the DBEs.

ASSIGNMENT

1. What are the main components of a plant protection system?
2. Why is it necessary to consider the response time of instruments, safety logic, and shutdown system while arriving at the limiting safety system settings?
3. Why do we need a trip threshold on the reactivity signal in a SFR unlike a PWR?
4. What is the need to calculate the permissible delay time for trip action on different trip parameters? How is this accomplished?

REFERENCES

AERB Document, (1990), *Safety Criteria for the Design of Prototype Fast Breeder Reactor*, Atomic Energy Regulatory Board, India.

APDA-233, (1966), Report on the fuel melting incident in the Enrico fermi atomic power plant on October 5, 1966, doi:10.2172/4766757, https://www.osti.gov/servlets/purl/4766757

IAEA Study Series-246, (1985), *Status of Liquid Metal Cooled Fast Breeder Reactors*, IAEA, Vienna.

Misra M.K., Sridhar N. and Murthy D.T. (2014), "Design and Implementation of Safety Logic with Fine Impulse Test System for a Nuclear Reactor Shutdown System," *2014 27th International Conference on VLSI Design and 2014 13th International Conference on Embedded Systems*, pp. 198–203, doi:10.1109/VLSID.2014.41.

Natesan K. et al., (2012), Plant dynamics studies towards design of plant protection system for PFBR, *Nuc. Eng. Design*, Vol. 250, pp. 339–350; doi:10.1016/j.nucengdes.2012.05.009

Senthil Kumar C., John Arul A., Pal Singh O. and Suryaprakasa Rao K. (2005), Reliability analysis of shutdown system, *Ann. Nucl. Energy*, Vol. 32, pp. 63–87.

Vaidyanathan, G., Sangodkar, D.B. and Paranjpe, S.R. (1987), Limiting safety system settings for FBTR operation, Specialists' meeting on Load following control of nuclear power plants including availability aspects, International Atomic Energy Agency, Vienna (Austria). International Working Group on Nuclear Power Plant Control and Instrumentation; p. 243.

Chapter 12

Decay Heat Removal System

12.1 INTRODUCTION

Shutdown systems and decay heat removal systems form the backbone of the nuclear plant protection system. While the former ensures safe shutdown of the fission reaction, the latter is essential to remove the heat from the decay of the fission products during earlier fissions. Thus, heat continues to be generated even after shutdown. Residual or decay heat removal (DHR) systems are needed to ensure that fuel clad temperatures do not rise. In addition to fission product decay, some decay heat is produced by beta decay of U_{239} and Np_{239} plus smaller amounts from decay of activation products (e.g., steel, sodium) and higher-order actinides such as Cm_{242}. Also, in situations of loss of off-site and on-site power, this heat needs to be removed. Toward this, one has to ensure that the design is amenable to natural circulation cooling in different situations. The primary coolant system in modern SFRs can easily be configured to provide natural circulation shutdown heat removal. The capability to remove shutdown decay heat with natural circulation provides a means to maintain reactor component temperatures at acceptable levels even in the event of loss of all off-site and on-site power supplies. This chapter looks at the different decay heat removal options, modeling of such systems, and obsevations based on tests and studies done in different countries.

12.2 NATURAL CONVECTION BASICS

Natural circulation flow arises due to the effect of gravity on a continuous fluid with a density difference along the elevation. Heavy fluid sinks to displace lighter fluid. Buoyancy-induced flow can be established when a fluid is heated, decreasing its density, at an axial position below the elevation at which the fluid is cooled (increasing its density). In a one-dimensional scenario, flow occurs when the buoyancy force is great enough to overcome form, friction, and shear losses. The natural circulation flow rate is regulated by the balance between the buoyancy force and the flow-related pressure losses.

DOI: 10.1201/9781003283188-12

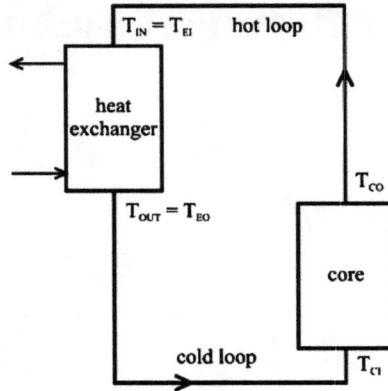

Figure 12.1 Simple representation of reactor circuit.

When the buoyancy force is provided by a thermally driven density difference, the fluid flow rate will be determined by the fluid properties, the elevation difference between the heat sink and the heat source, and the density difference in the fluid between the heat source and the heat sink as caused by the heating and cooling. The flow rate and hot-to-cold coolant differential temperature relationship of single-phase natural circulation in steady-state conditions in a closed circuit (Figure 12.1) is simple to derive as shown by Pavel and Vaclav (2014):

$$\dot{m} = \left[\frac{2\bar{\rho}g\beta P}{\bar{k}c_p} \left(\overline{z}_{hx} - \overline{z}_c \right) \right]^{1/3}$$

$$\Delta T_c = \left(\frac{P}{\bar{\rho}c_p} \right)^{2/3} \left[\frac{\bar{k}}{2g\beta \left(\overline{z}_{hx} - \overline{z}_c \right)} \right]$$

where:
    $\dot{m}$ = total natural circulation flow rate,
    ΔT_c = core outlet minus core inlet temperature differential,
    $\bar{\rho}$ = average coolant density in the core,
    g = gravitational force,
    β = volumetric coef. of thermal expansion of sodium,
    P = power added/removed in the core/heat exchanger,
    $\bar{k}$ = total pressure loss per area^2 around the loop,
    c_p = specific heat for sodium,
    $\overline{z}_{hx}$ = heat exchanger midplane elevation,
    $\overline{z}_c$ = core midplane elevation.

This approach can be used in the preliminary design to fix the elevation differences between the source and the sink in decay heat removal systems.

Liquid sodium and its alloys are excellent fluids for natural circulation heat removal because of their thermophysical properties. Due primarily to its high thermal conductivity, liquid sodium is capable of very high convective heat transfer rates, even at the modest fluid velocities, characteristic of natural circulation. This tends to minimize the temperature differences between the heat source and the fluid, and between the fluid and the heat sink, and to reduce the overall source-to-sink temperature difference required for natural circulation cooling.

12.3 DHR SYSTEM OPTIONS IN SFR

The key design parameters to ensure natural circulation heat removal in SFR systems are (1) provision for a relatively free-flowing fluid natural circulation path and (2) provision for sufficient elevation difference between the heat source and the heat sink.

12.3.1 DHR in Primary Sodium

In the primary coolant circuit, natural circulation flow may be established along the same flow path used for normal operation. Along this path, coolant is heated in the reactor, rises to the hot plenum, and flows through the intermediate heat exchangers (IHX) to the cold plenum and back to the reactor core. In accidents or emergency shutdown conditions in which no heat is removed in the IHX, heat can be removed by independent heat exchangers in series with the IHX for loop-type reactor designs. For pool-type reactor designs, this function is carried out by the direct reactor auxiliary cooling system (DRACS) where dip heat exchangers may be located high in the hot pool (Figure 12.2). DRACS is also referred to as a safety grade decay heat removal system (SGDHR) to convey its importance to safety. Heat removed from the hot pool through a decay heat exchanger is transferred to the outside envionment through an air-cooled heat exchanger.

A variation to the above would be a similar decay heat removal system in the cold pool. Primary coolant cooled in the DRACS heat exchangers falls to near the bottom of the cold pool, where it enters the primary coolant pump inlet and travels back to the reactor. However, this system would not be effective in case of an accident involving core melting, when the flow path through the core would be obstructed.

12.3.2 DHR in Secondary Sodium

For either loop- or pool-type reactor configurations, primary coolant natural circulation carries heat from the reactor to the IHX or auxiliary heat

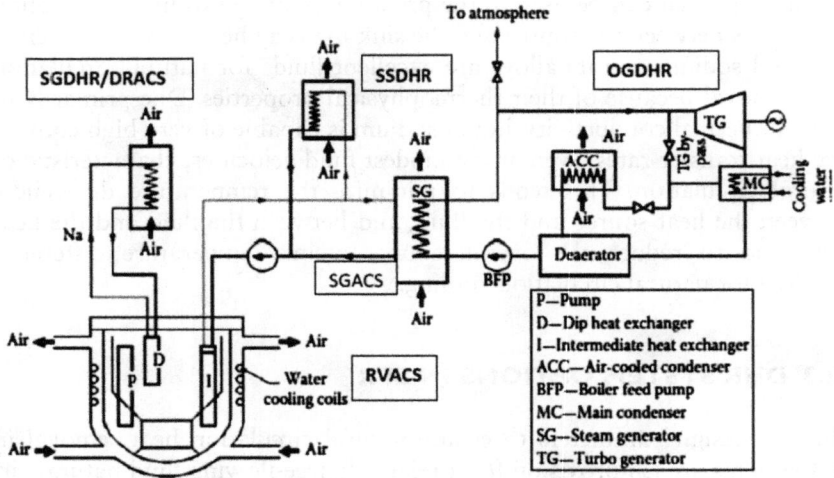

Figure 12.2 Decay heat removal system in pool, secondary sodium, and steam water circuits

Source: Baldev Raj et al., *Sodium Fast Reactors in Closed Fuel Cycle*, CRC Press, 2015.

exchangers. Heat transferred at the IHX may be removed by natural circulation in the secondary sodium loop, refered to as secondary sodium decay heat removal system (SSDHRS), if sufficient elevation of the ultimate heat sink is provided (Figure 12.2).

12.3.3 Steam Generator Auxiliary Cooling System

A steam generator auxiliary cooling system (SGACS) involves the placement of SG modules in casings with dampers that are opened to allow atmospheric air, and heated air is let out through a tall chimney on the top of the casing to provide sufficient draft (Figure 12.2).

12.3.4 DHR Through Steam Water System

After a reactor trip, the turbine is not available, and steam coming from SG bypasses the turbine and is led to an air-cooled condenser. This is referred to as an operational-grade decay heat removal system (OGDHR). This consists of a steam water separator tank, steam–air decay heat condensers, and recirculating pump along with associated circuit that carry heat from a steam water system to the ambient air (Figure 12.2). Following SCRAM, when the temperature of steam produced in the SG reduces below that of saturation, steam is automatically directed through the moisture separator tank, and the water level in the tank starts building up. This system is operable only when the secondary sodium system, steam water system, and class 4 power supply are available.

12.3.5 Reactor Vessel Auxiliary Cooling System

With reactor vessel auxiliary cooling system (RVACS), the decay heat is transferred from the reactor core to the cold pool by natural circulation of sodium coolant. The heat is then conducted through the reactor vessel wall and transferred across an argon/nitrogen gap by radiation to the containment vessel (Figure 12.3). The heat is further conducted through the containment vessel and then removed from outside of the containment vessel by natural circulation of upwardly flowing atmospheric air around the vessel. The rate of heat removal is controlled primarily by the radiation heat transfer through the argon/nitrogen gas from the reactor vessel to the guard vessel. The radiation heat transfer is a strong function of emissivity of the reactor vessel and safety vessel material. This may change with time depending on local environment conditions.

In a variation to air cooling, in the french reactors PHENIX and Super PHENIX, tube panels with water inside reactor pit walls and bottom receive

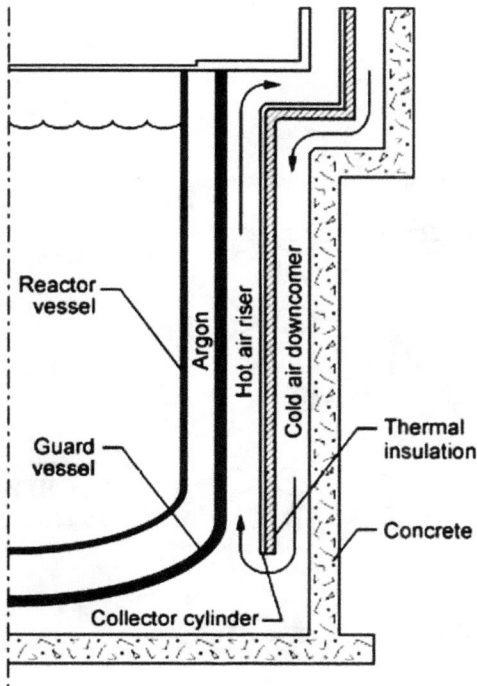

Figure 12.3 Reactor vessel auxiliary cooling system.

the radiated heat from the safety vessel and participate in decay heat removal. Experience in the Super PHENIX reactor showed that while for initial design, the reactor vessel emissivity coefficient was considered as being 0.6, measurements on vessel austenitic steel specimens were 0.4 at 500°C and 0.5 at 700°C (Joël and Gérard, 2017). Later, the balances of the power removed, measured during isothermal tests at 395°C, corresponded to a vessel emissivity coefficient theoretical estimate even lower, of 0.25. Hence, the heat removed is a strong function of the emmissivities of the steels used and the environmental conditions to which they are exposed.

12.4 DHR IN FBTR

FBTR is a loop-type reactor with two primary sodium loops feeding a core in the reactor vessel, two secondary sodium loops, two intermediate heat exchangers (IHX), and four modules of serpentine-type SGs along with the steam/water circuit and a turbine generator system (Srinivasan et al., 2006). The four modules of the SG are enclosed in an insulated casing that is provided with four inlet trap doors and a 6.6-m-tall stack. The schematic of the SG casing and air stack is shown in Figure 12.4.

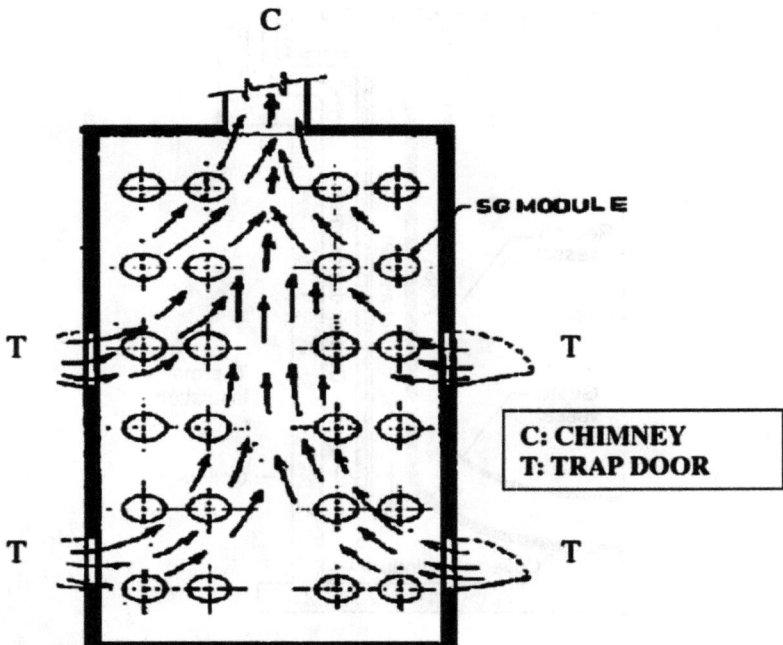

Figure 12.4 FBTR SG modules in casing.

Table 12.1 Plant Conditions and DHR in FBTR

S. No.	DHR path	Conditions for DHR by this path	Event
1	Normal heat transport	Off-site power, at least one secondary circuit, and steam water system available with dump condenser	Reactor trip, Turbine trip
2	SG air cooling and piping losses (~1 MWt)	Both primary and secondary systems with SG casing intact	Loss of off-site and on-site power failure, Steam water system not available
3	Reactor vessel nitrogen preheating circuit (~350 KW)	Availability of on-site power supply	Sodium leaks secondary circuits
4	Biological shield cooling system (~350 KW)	Availability of on-site power supply	Sodium leak in secondary circuit, Primary sodium leak outside the reactor vessel in piping

The various means of DHR available in FBTR are: (1) normal heat transport path, i.e., primary sodium, secondary sodium, and steam/water circuits; (2) heat losses from primary and secondary sodium systems and heat removal from SG casing; (3) forced circulation of nitrogen in the preheating and emergency cooling circuit (PHEC) connected to the double envelope of the primary sodium circuit and reactor vessel; and (4) forced convection water cooling of the biological concrete shield.

Table 12.1 offers a brief summary of the plant conditions required for successful DHR by the above cooling modes, their capacity, and the situation for which these modes are put in service.

In the event of non-availability of the steam/water system, it is envisaged to effect DHR through the SG casing. In order to substantiate safe DHR through this mode, static and transient calculations have been carried out as follows.

12.4.1 Heat Removal by Air in SG Casing

For steady-state calculations, it was considered that heat is convected by air flow over the SG shell and the SG casing wall inner surface. Heat from SG shell to casing wall is transferred by radiation. Air flow is established by matching the buoyancy head with the frictional pressure drop due to air flow. The solution of the equations is carried out in an iterative manner:

$$Q = W_{air}Cp\left(T_{air}^{hot} - T_{air}^{cold}\right)$$

where W_{air} is the air flow in SG casing, Cp_{air} is the specific heat capacity of the air, and T_{air}^{cold} and T_{air}^{hot} are, respectively, the temperatures of air entering and exiting each SGs casing.

W_{air} can be obtained by equating the driving head for air with the pressure drop in the casing as the air flow sover SG shell:

$$\Delta P = \Delta \rho g H_{cas} = \beta_{air} \left[T_{air}^{hot} - T_{air}^{cold} \right] g \, H_{cas} = k_{casing} W_{air}^2$$

where H_{cas} is the height of the casing and K_{casing} is the pressure drop coefficient for air flow over SG shell in the casing. From the above equation we get

$$W_{air} = \gamma_{SGcasing} \sqrt{(T_{air}^{hot} - T_{air}^{cold})}$$

$$\gamma = \sqrt{\beta_{air} g H_{cas} / k_{casing}}$$

where β_{air} is the linear coefficient linking density variation to temperature variation.

Using the log mean temperature difference approach we can write the equation of heat transferred from sodium to air,

$$Q = h_{cas} S_{cas} \left[\frac{\left(T_{na}^{hot} - T_{air}^{hot} \right) - \left(T_{na}^{cold} - T_{air}^{cold} \right)}{\log_n \left(\dfrac{T_{na}^{hot} - T_{air}^{hot}}{T_{na}^{cold} - T_{air}^{cold}} \right)} \right]$$

where h_{cas} is the overall *heat* transfer coefficient, S_{cas} is the exchange surface area, and T_{na}^{cold} and T_{na}^{hot} are, respectively, the temperatures of the secondary sodium fluid entering and leaving the SG.

Also,

$$Q = W_{na} Cp_{na} \left(T_{na}^{hot} - T_{na}^{cold} \right)$$

$$Nu_{air} = C \left(GrPr \right)^n$$

where $C = 0.59$ and $n = 0.25$ for $10^4 < GrPr < 10^9$
and

$C = 0.021$ and $n = 0.4$ for $10^9 < GrPr < 10^{15}$ [Morgan, 1997]

The air side heat transfer coefficient will be the controlling one as compared to sodium. Sodium heat transfer coefficient can be calculated based on the following correlation:

$$Nu = 0.625 Pe^{0.4}$$

The air flow and temperature of sodium at SG outlet are varied until the conditions of energy balance and equality between the driving force and pressure drop on the air side are obtained. The radiative heat transfer from the SG shell to the SG casing wall and the heat convected by the air flowing over the casing walls need to be considered. The temperature of the inner surface of the casing is calculated in an iterative manner by matching the radiative heat transfer to the casing and convective heat removal over the inner surface of the casing by the air flow. Studies were conducted for various sodium inlet temperature to the SG, sodium flows of 50 kg/s and 2 kg/s representing, respectively, forced and natural convection sodium flow regime in the SG, and either all four inlet doors are effective or only two of them are. Studies indicated a heat removal capability of 258 KW and 239 KW, respectively, at 215°C for four trap door and two trap door opening (Kasinathan et al., 1993). For a sodium temperature of 350°C, the corresponding figures are 549 KW and 512 KW. Tests performed in FBTR for the forced sodium flow conditions have shown that heat removal is 260 KW at 215°C and 580 KW at 350°C, respectively, for the four trap door opening case.

12.4.2 Loss of Off-site and On-site Power with SG Air Cooling

Loss of off-site and on-site power leads to a situation often called *station blackout* (SBO). An SBO event is simulated by considering the trip of all primary and secondary sodium pumps along with instantaneous loss of feedwater flow through SG. The emergency battery power backup provided for the operation of primary pumps at a low speed (150 rpm) for a short duration of 1/2 h is not considered conservatively. In this case, off-site power failure leads to SCRAM. This event is analyzed by considering reactor SCRAM at 6 s by the high core outlet temperature parameter, when the reactor is operating at a power of 22.1 MW (55% of full power). Analysis was carried out with DYNAM code with suitable additional module to represent SG casing heat removal (Natesan et al., 2013).

The analysis indicated that reactor power falls to decay power level within 7 s and reduces subsequently. Decay power values after 1 day and 10 days are 135 kW and 80 kW, respectively. SG trap doors are manually opened at 1800 s (delay for manual action). They are closed manually when the sodium temperature at the outlet of SG (TSGO) reaches 200°C, and they are again considered to be opened when TSGO reaches 260°C. Short term and long-term evolutions of clad hotspot temperature are shown in Figure 12.5. Two peaks are observed in the clad hot spot temperature with the first peak of 704°C occurring close to the SCRAM instant and the second peak of 675°C occurring during the evolution of natural convection. The second peak is due to the closure of trap doors when TSGO falls below 200°C around 1 h.

Figure 12.6 shows short- and long-term evolutions of sodium flows (per loop) in the primary and secondary sodium circuits. Oscillations with a

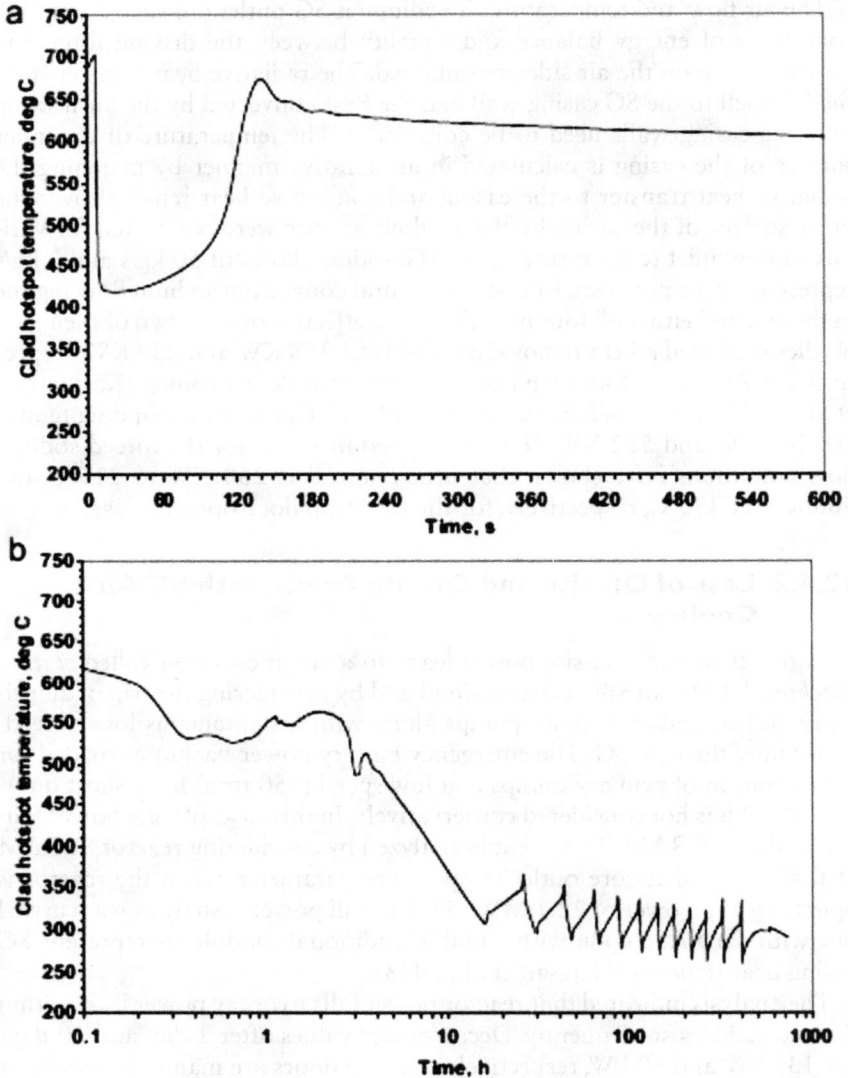

Figure 12.5 (a) Short-term and (b) long-term evolution of fuel clad hot spot temperature.

Source: Natesan K. et al., Thermal Hydraulic Investigations of an Extended Station Blackout Event in FBTR. *Nuclear Engineering and Design* 265, 2013.

smaller period are observed during the initial duration of the transient (up to 4 h) before a reasonably stable natural convection flow is established. It can be observed that whenever the trap doors are opened, secondary sodium flow oscillates before stabilization. This is due to the presence of the surge

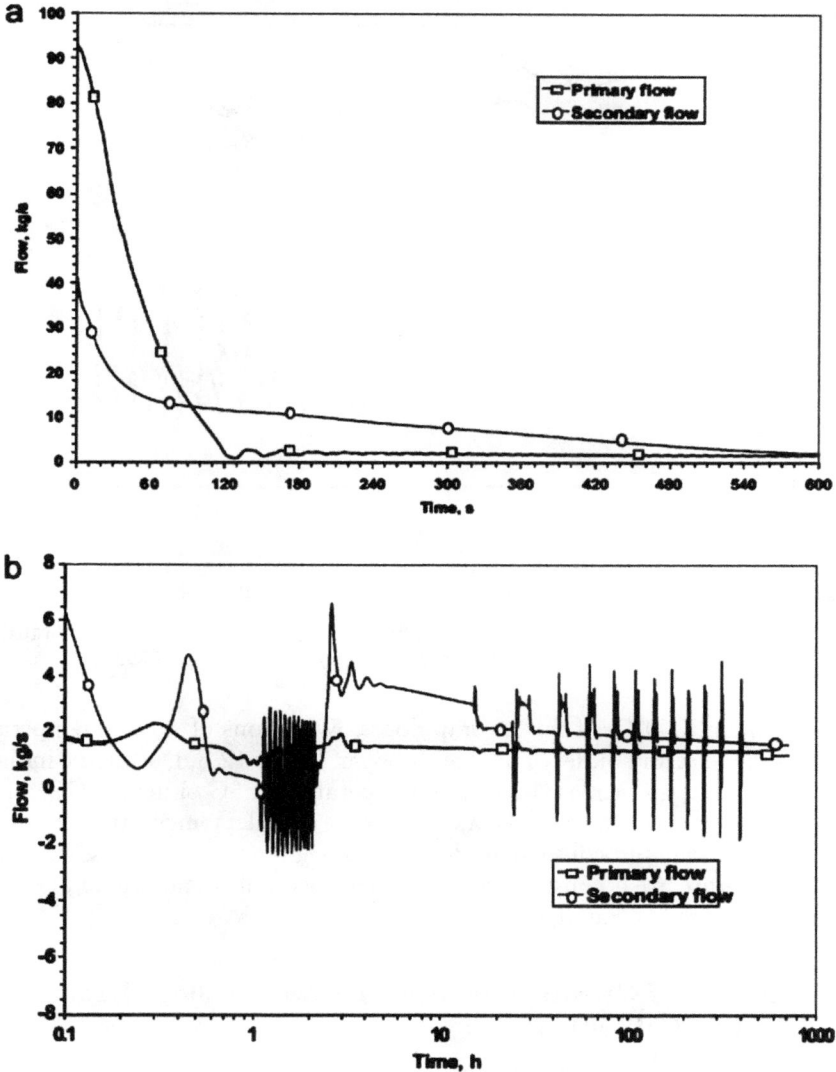

Figure 12.6 (a) Short- and (b) long-term evolution of sodium flows.

Source: Natesan K. et al., Thermal Hydraulic Investigations of an Extended Station Blackout Event in FBTR. *Nuclear Engineering and Design* 265, 2013.

tank and the pump tank at the same high elevation in the hot and cold leg piping, respectively. These tanks delay the propagation of relatively faster thermal transient occurring in SG (due to trap door opening) to the IHX side.

Long-term evolutions of hot pool and reactor inlet temperatures are shown in Figure 12.7. The oscillations are essentially due to closing (200°C)

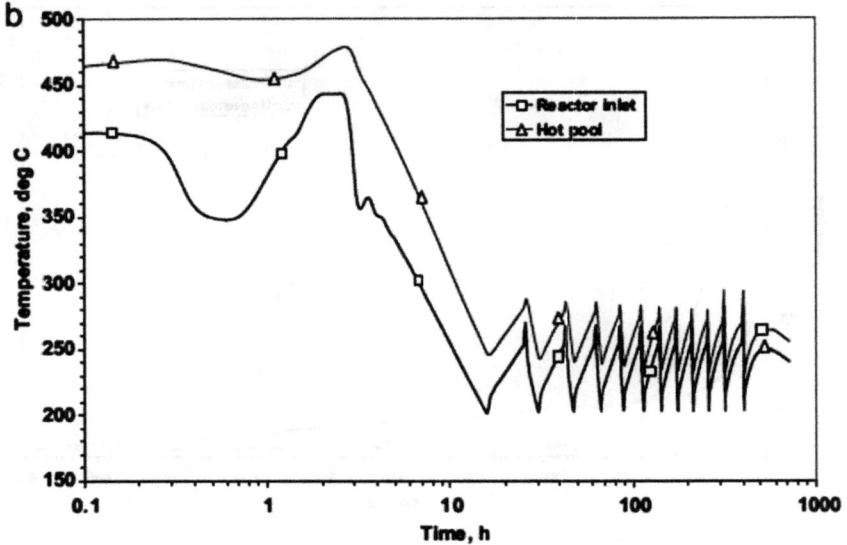

Figure 12.7 Long-term evolution of reactor inlet and hot pool temperatures.

Source: Natesan K. et al.,Thermal Hydraulic Investigations of an Extended Station Blackout Event in FBTR. *Nuclear Engineering and Design* 265, 2013.

and opening (260°C) of the SG trap doors. Evolutions of sodium tempera-ture at the inlet and outlet of SG are shown in Figure 12.8. During the initial period, there is an increase in sodium temperature at SG outlet (TSGO) due to loss of heat sink. Later it follows the drop in the inlet temperature because of a reactor trip and consequent drop in hot leg temperatures. Oscillations in temperature are a consequence of oscillations in flow, and these are due to operation of SG trap doors.

12.4.3 Loss of Off-site and On-site Power without Reactor Trip (ULOF)

This is a low-probability event, as it requires simultaneous loss of off-site power and on-site power (station diesels, batteries) and failure of safety logic to trip the reactor (Vaidyanathan et al., 2010). In view of the fact that FBTR has only one shutdown system comprising control rods, a diverse shutdown system is absent unlike in present-day fast reactors that have two redundant and diverse shutdown systems, which will ensure shutdown with high reliability. Hence the interest to study the case of loss of off-site and on-site power considering the failure of a shutdown system. Main decay heat removal paths in the case of loss of off-site and on-site power in FBTR are the piping losses and heat removed by natual convection of air over the SG surfaces placed in the casing. For such extreme events

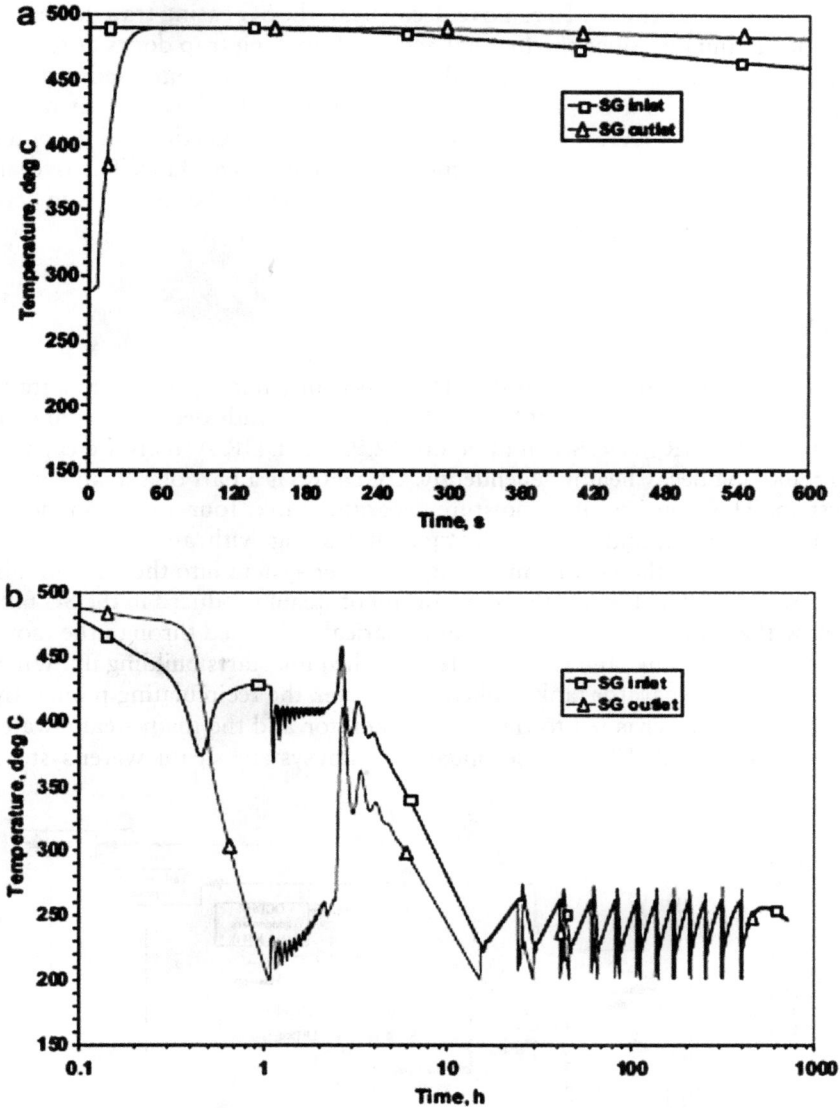

Figure 12.8 (a) Short-term and (b) long-term evolution of SG temperatures

Source: Natesan K. et al., Thermal Hydraulic Investigations of an Extended Station
    Blackout Event in FBTR. *Nuclear Engineering and Design* 265, 2013.

one needs to respect the limit on the bulk sodium temperature (Gouriou
et al., 1982). Studies were carried out at an extended power of 45 MWt
with both the primary and secondary pumps coast down as governed by
their drive inertia and the boiler feed pumps stopping supply of water

almost instantaneously. To remove decay heat, the SG casing trap doors are opened manually. Since the manual action of opening trap doors at the SG site involves a delay, a maximum delay time of half an hour is considered in the analysis as a conservative measure. The study showed that maximum sodium temperature in the core reaches ~930°C (the boiling point of sodium) and in less than a minute it drops as power falls. The overall temperatures stabilize around 475–525°C. Details can be found elsewhere (Vaidyanathan et al., 2010).

12.5 DHR IN PFBR

In PFBR, there are two parallel DHR systems, namely operation-grade decay heat removal system (OGDHRS) and safety-grade decay heat removal system (SGDHRS), as shown in Figure 12.9. The DHR systems are capable of removing decay heat independently. OGDHRS is a part of a steam–water system. This consists of a moisture separator tank, four steam–air decay heat condensers, and a recirculating pump along with an associated circuit that pushes the heat from the steam–water system into the ambient air. Following SCRAM, when the temperature of steam produced in the SG falls below that of saturation, steam is automatically directed through the moisture separator tank, and the water level in the tank starts building up. When the moisture separator tank is filled with water, the recirculating pumps are started, and water is fed to the steam generator and the main steam–water system is isolated. When the secondary sodium system, steam–water system,

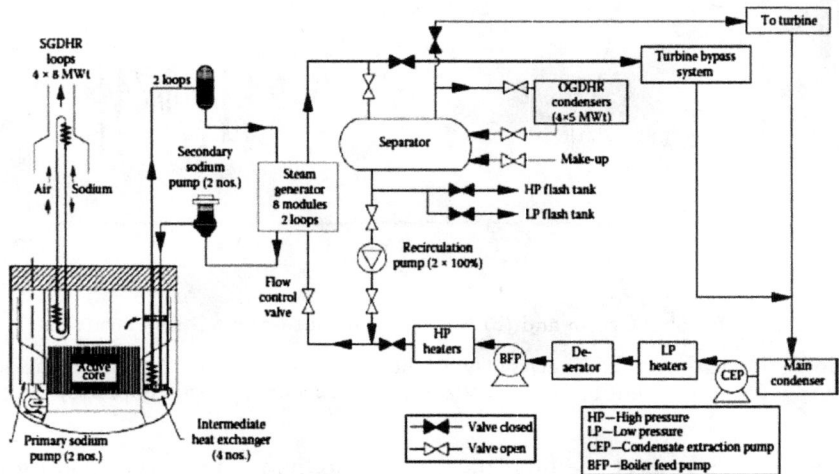

Figure 12.9 Schematic of decay heat removal in PFBR.

Source: *Baldev Raj et al., Sodium Fast Reactors in Closed Fuel Cycle*, CRC Press, 2015.

and class 4 power supply are available, the DHR operation is carried out by OGDHRS. But dependence of the OGDHRS on class 4 power supply makes the system less reliable.

Whenever the OGDHRS is not available, the DHR is carried out by more reliable SGDHRS. SGDHRS consists of a sodium–sodium heat exchanger (DHX) and a sodium–air heat exchanger (AHX). The DHX is dipped in the hot pool, and the AHX is located on top of the steam generator building, which is at an elevation of ~42 m above the DHX.

An intermediate sodium circuit connects the DHX and AHX (Figure 12.10). The AHX is placed inside a well-insulated casing. A tall stack of 30 m in height is provided on top of the AHX. The DHX removes decay heat directly from the hot pool, and the AHX pushes the heat out into the ambient air, representing the final heat sink.

The DHX primary side sodium flow, intermediate circuit sodium flow, and air flow over the AHX are by natural convection. The locations of the DHX and the AHX provide a sufficient buoyancy head for the intermediate

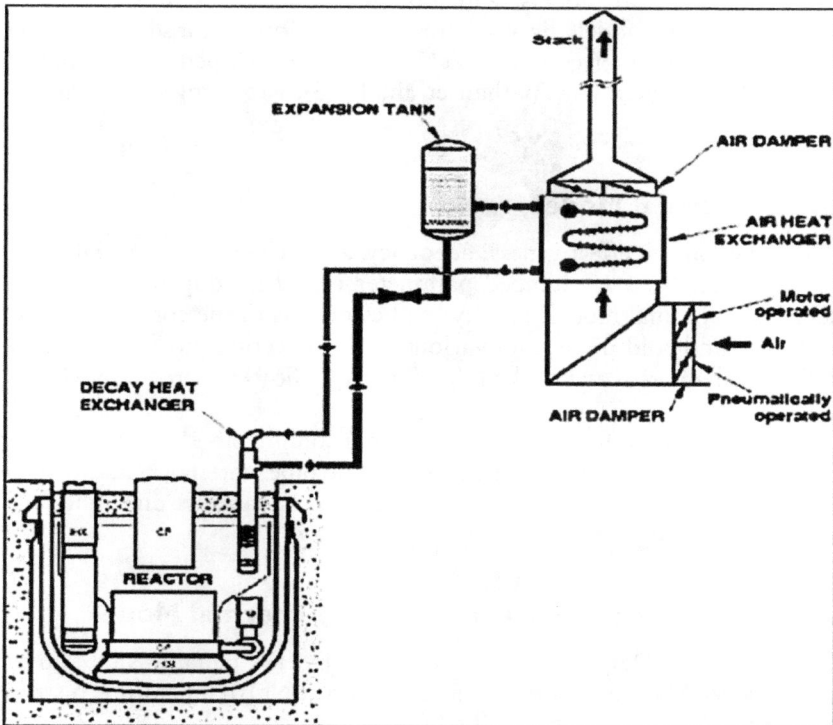

Figure 12.10 Schematic of a SGDHR circuit of PFBR.

Source: Baldev Raj et al., *Sodium Fast Reactors in Closed Fuel Cycle*, CRC Press, 2015.

circuit sodium flow, and the AHX stack provides natural draft for air flow in the AHX. This makes the SGDHRS passive except for the opening of AHX air dampers. The AHX is provided with two dampers at the inlet and two at the outlet. One damper at the inlet and one damper at the outlet are operated by class 1 power supply, and the other two dampers are operated pneumatically. On demand, the signals go to the damper-opening mechanisms automatically. In case of signal failure, the operator can open these dampers remotely from the control room or manually from the location of the dampers using a hand wheel. During power operation of the reactor, the dampers are kept in crack-opened position to maintain a small air flow to ensure that natural convection takes place in the proper direction.

For a system code, it is necessary to model the SGDHR system and couple it suitably to the main plant model to arrive at realistic predictions of plant temperatures. One of the main objectives is to ensure that plant temperatures at different points are within limits. At the design stage, the parametric studies required are large to finalize the layout of SGDHR. At the final stage after completion of the design with one-dimensional model, use of multidimensional models, with boundary conditions drawn from the one-dimensional models, is used for confirmation of the design. A one-dimensional computer code DHDYN has been developed for the analysis of a SGDHR system (Kasinathan et al., 1993). Modeling details are discussed below.

12.5.1 Thermal Model

To understand the effectiveness, adequacy, and behavior of a SGDHR system, traditionally a 1D lumped parameter model is adopted. This should include the thermal models and hydraulics models of the core, IHX, DHX, AHX, hot and cold pools, and various interconnecting pipes. As discussed earlier in Chapters 3 and 4, the following simplifications are adopted:

- Liquid sodium flow is assumed to be incompressible and single phase.
- Axial conduction heat transfer in coolant, pipe material, heat exchanger tubes, fuel pin, and clad walls is neglected in comparison to the radial conduction.

12.5.2 Decay Heat Exchanger (DHX) Thermal Model

The modeling of DHX is similar to that of IHX. The equations that govern the primary (PNa) and intermediate sodium (SNa) temperature distributions are

$$C_{PNa}\left(\frac{dT_{PNa}}{dt}\right) = \left(WC_P\right)_{Na}\left[\frac{\partial T_{PNa}}{\partial Z}\right] + \left(UA\right)_{PNa-SNa}\left[\left(\overline{T}_{SNa} - T_{PNa}\right)\right]$$

$$C_{SNa}\left(\frac{dT_{SNa}}{dt}\right) = \left(WC_P\right)_{SNa}\left[\frac{\partial T_{SNa}}{\partial Z}\right] + \left(UA\right)_{PNa-SNa}\left[\left(\overline{T}_{PNa} - \overline{T}_{SNa}\right)\right]$$

where T, W, Cp, U, and A refer to temperature, flow, specific heat, overall heat transfer coefficient, and heat transfer area, respectively. Subscripts p and s refer, respectively, to primary and intermediate sodium.

The Nusselt number for sodium is estimated from

$$Nu = 0.625Pe^{0.4}$$

12.5.3 Hot Pool Thermal Model

Hot primary sodium enters DHX from the top portion of the hot pool. Cold primary sodium exiting from DHX mixes with hot pool sodium at the bottom of the hot pool. Modeling the hot pool and cold pool as single mixing volumes does not bring out the effects of thermal stratification and the plume effect of a relatively hot flow into a cold plenum, or vice versa. These effects can be better modeled by subdividing the hot or cold pool into two or more discrete volumes and specifying the plenum inlet flows to take appropriate path depending on the inlet and the plenum temperatures.

A three-zone model of a hot pool has been considered, namely one zone comprising sodium above the DHX primary inlet bottom, another below the DHX primary outlet window top, and the third one that lies between the other two zones and is in contact with IHX inlet windows (Figure 12.11). This approach is based on the natural circulation experiments carried out in scaled-down water model for the Super PHENIX reactor (Asteigno et al., 1981).

The flow path of sodium in the zones is based on a net pressure gradient that involves the mixed mean temperature of each zone. The temperature distributions are used to calculate the buoyancy induced forces. The thermal process occurring in each zone can be represented by the following equations for the flow path represented at the top in Figure 12.11. Equations for the case of reactor outlet sodium being lower than the hot pool zone temperatures, which is the case immediately after a reactor trip, is given below.

Temperature of bottom zone I is given by

$$\left(MC\right)_I \frac{dT_I}{dt} = \sum W_i T_i + W_D T_{Do} - WT_I + hA\left(T_{II} - T_I\right)$$

$$W = \sum W_i$$

where MC is the mass capacity of top zone I as indicated in the figure, W_i, T_i are, respectively, the flow and temperature of various reactor core

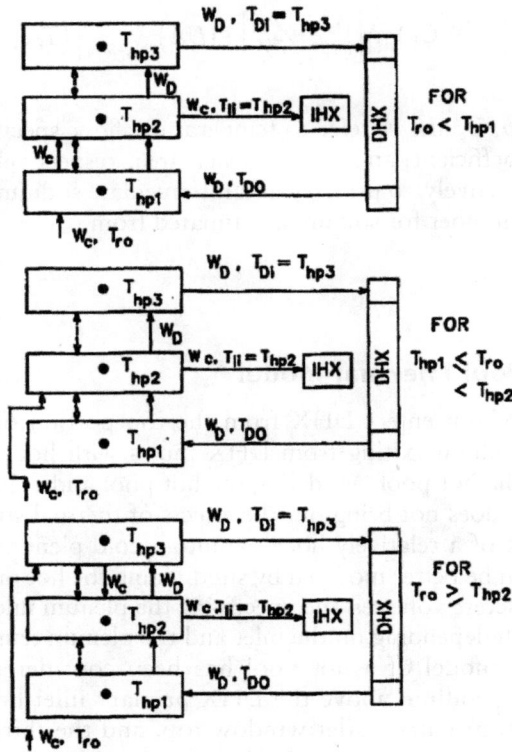

Figure 12.11 A three-zone model of a hot pool

Source: Kasinathan N. et al., Evaluation of DHR studies for Indian FBR program, IAEA Specialists Meeting, Oarai, Japan, February 1993.

flow zones, W_D is the DHX outlet flow into hot pool zone I, and T_{DO} is the temperature of primary sodium at the DHX outlet. The last term in the equation represents heat transfer between zone II and zone I of the hot pool.
   Temperature of middle zone II is given by

$$(MC)_{II} \frac{dT_{II}}{dt} = WC_p \left(T_I - T_{II} \right) - hA \left(T_{II} - T_{III} \right) - hA \left(T_{II} - T_I \right)$$

where the first term is the heat transported into zone II from zone I, the second term represents the heat transfer between hot pool zones II and III, and the last term represents the heat transfer between hot pool zones II and I.
   Temperature of top zone III is given by

$$(MC)_{III} \frac{dT_{III}}{dt} = WC_p \left(T_{II} - T_{III} \right) + hA \left(T_{II} - T_{III} \right)$$

where the first term is the heat transported into zone III from zone II and the second term represents the heat transfer between zones II and III.

The heat transfer between the different hot pool zones is based on conductive heat exchanges between the zones. Similar equations can be written for the middle and bottom flow paths depicted in Figure 12.11. The cold pool is also treated in a three-zone model, same as the hot pool. The overall computational model is shown in Figure 12.12.

12.5.4 Air Heat Exchanger (AHX) Thermal Model

AHX is a cross-flow heat exchanger with sodium flowing inside finned tubes (Figure 12.13). Air flows from bottom to top, while sodium flows from top to bottom through serpentine tubes (Vinod et.al., 2013). To minimize air flow bypassing the tubes, the casing and tube supports are suitably designed. The air flow to the casing is controlled by suitable opening of the dampers. Since thermal capacity of air is very small compared to sodium, the accumulation term or inertial term in the air side heat transfer is neglected and a quasistatic model is used. Also, since sodium heat transfer coefficient is large, the temperature drops in the finned tube are small, and hence sodium and the finned tubes are treated as a single capacity.

The equations governing sodium and air temperatures in AHX are given below:

Sodium side:

$$MC_{na}\left(\frac{dT_{Na}}{dt}\right) = \left(WC_p\right)_{Na}\left[\frac{\partial T_{Na}}{\partial Z}\right] + \left(hA\right)_{Na-A}\left[\left(\overline{T}_A - T_{Na}\right)\right]$$

$$MC_A\left(\frac{dT_A}{dt}\right) = 0 = \left(WC_p\right)_A\left(\frac{\partial T_A}{\partial Z}\right) + \left(hA\right)_{Na-A}\left[\left(\overline{T_{Na}} - T_A\right)\right]$$

where MC, W, and h refer, respectively, to the mass capacity, flow, and heat transfer coefficient, and subscripts n and a, refer to sodium and air, respectively.

Heat transfer coefficient on the air side is the major controlling resistance to heat transfer. The following correlations are used on sodium and air side, respectively:

Nu = 4.82 + 0.0185 $Pe^{0.827}$

For sodium in the range of 58 < Pe < 1.31×10^4

For air side on finned tubes

$$Nu = 0.192\left(\frac{a}{b}\right)^{0.2}\left(\frac{5}{d_0}\right)^{0.18}\left(\frac{h_f}{d_0}\right)^{-0.14} Re^{0.65} Pr^{0.36}\left(\frac{Pr_f}{Pr_w}\right)^{0.25}$$

when 100 < Re < 2×10^4

Nu = 0.0507 when 2×10^4 < Re < 2×10^5

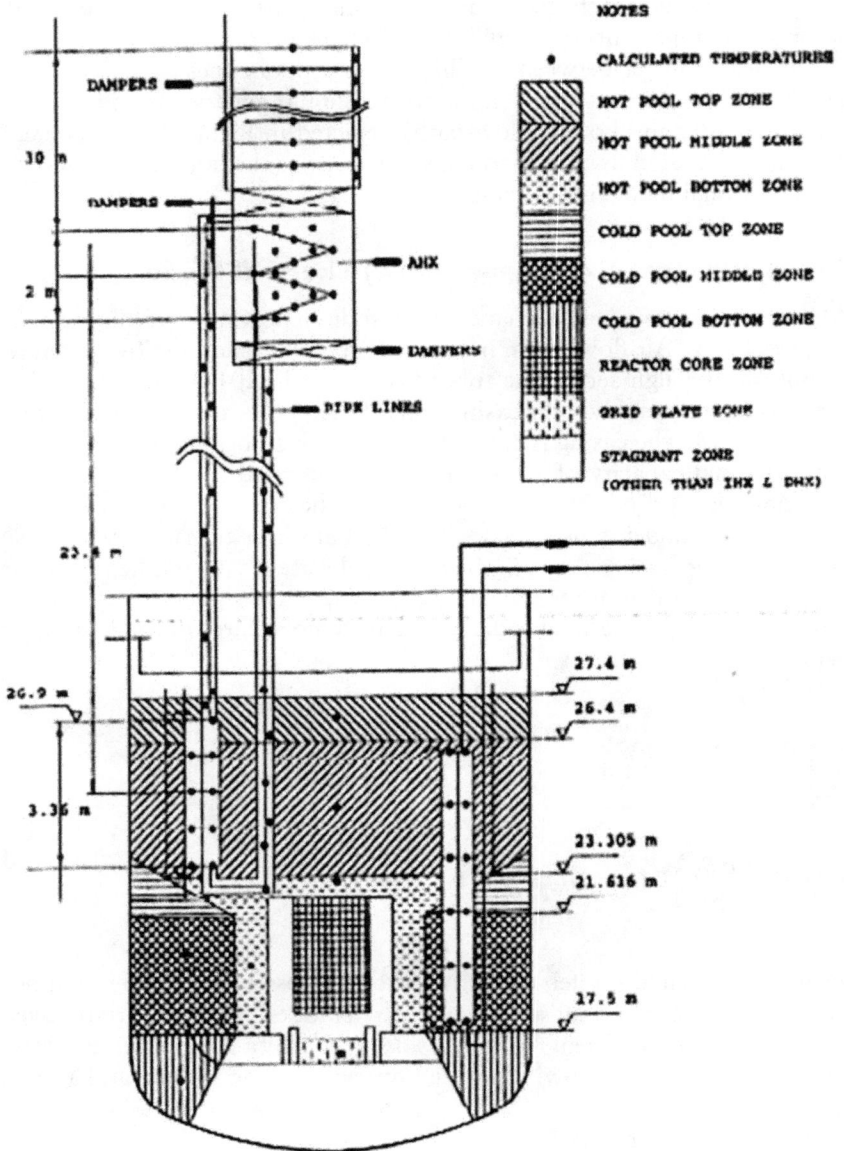

Figure 12.12 Computational model of a SGDHR circuit.

Source: Kasinathan N. et al., Evaluation of DHR studies for Indian FBR program,
         IAEA Specialists Meeting, Oarai, Japan, February 1993.

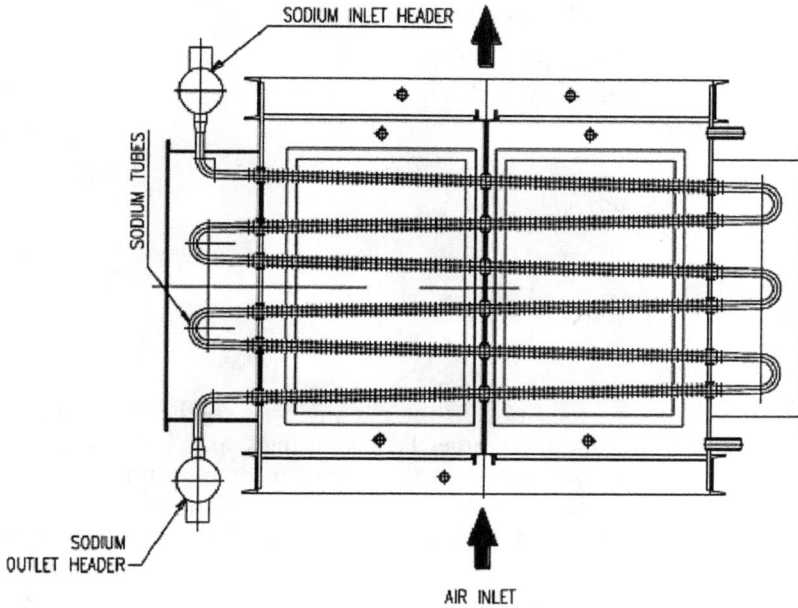

Figure 12.13 Sodium to air exchanger schematic.

Source: Vinod et al., Experimental Evaluation of Sodium to Air Heat Exchanger
Performance. *Annals of Nuclear Energy* 58, 2013.

where a and b are the relative transverse and longitudinal pitches, respectively, d_0 is the bare tube diameter, and h is the fin height (Schlunder, 1983).

12.5.5 Piping

The piping model is similar to the one used in Chapter 5.

12.5.6 Expansion Tank

The expansion tank in any sodium circuit is placed at the highest point to take care of thermal expansion of sodium as the temperature changes from 150°C to 550°C. It is modeled as a perfect mixing device.

12.5.7 Air Stack/Chimney

To achieve sufficient natural convection of air, a chimney has been provided at the AHX outlet. It is insulated to minimize heat losses. It is modeled in a similar way to the piping.

12.5.8 Hydraulic Model of SGDHR

For evaluating the DHX primary flow, the sodium flow in AHX, and AHX air flows, incompressible one-dimensional momentum equation as given below can be used:

$$\Sigma\left(\frac{L}{A}\right)\left(\frac{dW}{dt}\right) = g\oint \rho dz - \Delta P_f$$

$$\Delta P_f = \frac{W^2}{2\rho A}\left(f\left(\frac{\Delta x}{D_t}\right) + K\right)$$

where L/A is the inertia and K is the pressure loss coefficient taken from Idelchek (1993) for straight pipe, bends, fittings, and bellows. For the damper, the pressure loss coefficient is obtained from appropriate tests.

The resistance coefficient for air flow over the AHX tube bundle is calculated using the formula for a staggered tube arrangement (Schlunder, 1983):

$$K\,t.\ bundle = 5.4\left(d^*/d_e\right)^{0.3}\left(\text{Re}_{d*}\right)^{-0.25}C_z.Z \quad \text{in the range.}$$
$$\text{of } 2.2\times10^3 < \text{Re}_{d*} < 1.8\times10^5$$

where

$$d^* = \text{relative diameter } d^* = d\left[\frac{A_c}{A_t}\right] + \left[\frac{A_f}{A_t}\right]\left[0.785\left(D^2 - d^2\right)\right]^{0.5}$$

d = bare tube outside diameter, m
D = fin diameter, m
A_c = bare tube surface area per unit length, m²/m
A_f = fin surface area per unit length, m²/m
A_t = total surface area per unit length m²/m
de = is equivalent diameter, given by
de = 2 $[s(s_1 - d) - 2.\ s_4.\ h]/(2.\ h + s)$
s = fin spacing, m
s_1 = transversal pitch of the tube, m
s_4 = fin thickness, m
h = fin height, m

In the Reynolds number, Re_{d*}, calculation velocity u is the maximum inter-tube velocity calculated for plain tube banks,

$$\text{Re}_{d*} = ud^*/v$$

C_z is the correction factor that takes account of the number of rows of tubes in the bundle and is calculated as

$$C_z = 1/Z \sum_{z=1}^{z} c_z$$

$$c_z = 0.934 + \frac{0.355}{(Z - 0.667)} \text{ for } Z < 6; c_z = 1 \text{ for } Z \geq 6$$

where Z is the total number of rows in the bundle.

Density is to be calculated based on the temperatures obtained from the thermal model. Flow driven by buoyancy can be estimated by integrating the density along the length of the tubes in DHX and AHX and piping. A similar approach can be used for calculating buoyancy forces in the chimney.

12.5.9 DHDYN Validation on SADHANA Loop

Toward validating the one-dimensional code, the experiments carried out in SADHANA sodium rig are simulated and the numerical results are compared with the measured values. SADHANA is an experimental facility available at IGCAR, Kalpakkam, India for evaluating the performance of SGDHRS (Padmakumar et al., 2013). Figure 12.14 shows the schematic of the loop. The facility is a 1:22 scaled-down model of one of the two types of SGDHR circuits used in a PFBR (8 MW) with a designed capacity of 355 kW. The facility mainly consists of a sodium-to-sodium decay heat exchanger, sodium-to-air heat exchanger, a vessel containing a sodium pool, and associated piping. A stack is provided on the AHX in the air circuit for natural air circulation. Only the hot pool is represented, and a heater replaces the core.

The SADHANA facility has been simulated using the modules of the DHDYN code. While the predicted sodium flow rates were in good agreement with the measured values, there was a 1.25% difference in terms of air flow. A significant deviation in predictions was seen in AHX sodium outlet temperature. This may be because of complex geometry of AHX and large uncertainty in air side flow and heat transfer calculations. The power removal was underpredicted by the computational model to the extent of 2–6%. Earlier experiments were conducted on a bigger AHX model of 2.5-MW theoretical power and tested at the steam generator test facility by giving rated flow of sodium and air through forced convection, and a an underprediction to the extent of 6.4% was observed (Vinod et al., 2006).

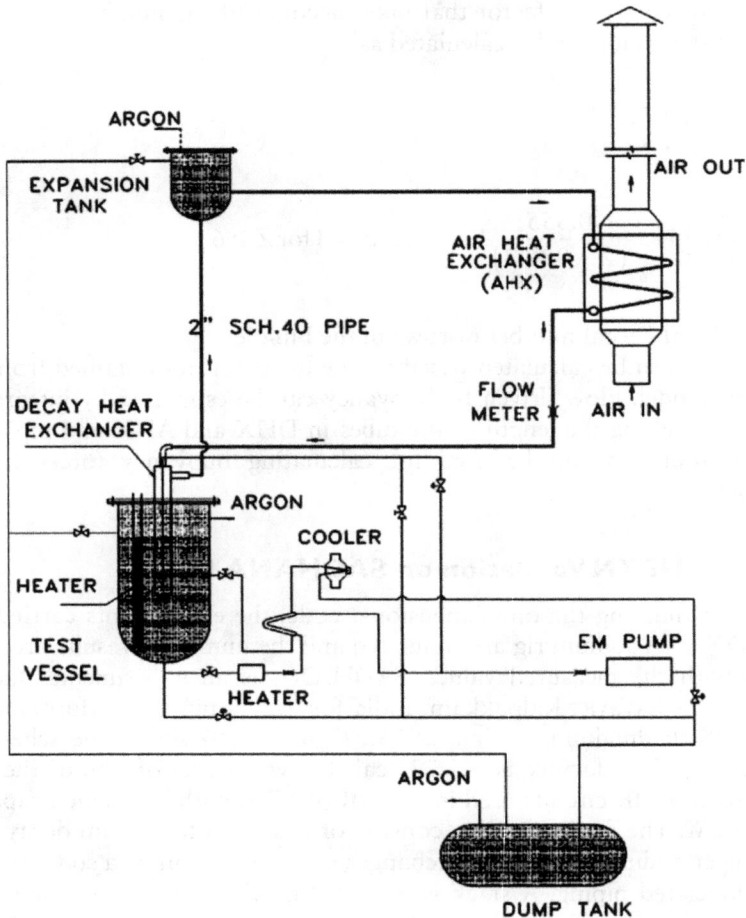

Figure 12.14 SADHANA natural convection loop schematic.

Source: Padmakumar G. et al., SADHANA Facility for Simulation of Natural Convection in the SGDHR System of PFBR. *Progress in Nuclear Energy* 66, 2013.

12.6 ROLE OF INTER-WRAPPER FLOW

During the loss of pumping power, natural convection currents are set up both inside the fuel SA and outside of it, in the inter-wrapper gaps between the SA. The former is supported by the buoyancy forces generated by the heat transfer from primary sodium to secondary sodium in the IHX, while the latter is due to the complex thermal hydraulic interaction among the inter-wrapper space and the hot pool. After a targeted burnup, spent fuel subassemblies (SA) are shifted to in-vessel storage locations. Such SA are called storage subassemblies (SSA). Since the power generated in SSA is

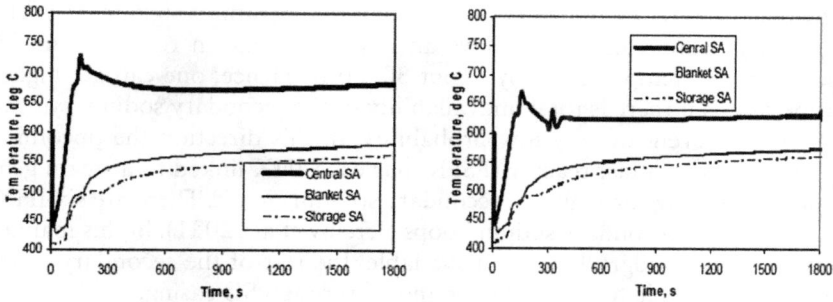

Figure 12.15 Evolution of subassembly outlet—loss of off-site power: impact of IWF.

Source: Parthasarathy U. et al., Estimation of Effect of Inter-Wrapper Flow Heat Transfer on Core Temperatures Using Porous Body Model, Proceedings of the 21st National and 10th ISHMT-ASME Heat and Mass Transfer Conference, December 27–30, 2011, IIT Madras, India.

much less than by fuel SA, the flow through SSA is reduced by providing extra pressure drop at the SA inlet.

With this arrangement it was seen that during DHR with the natural circulation flow in the primary circuit, the SSA clad hotspot temperatures exceeded allowable limits when the situation is analyzed by neglecting the inter-wrapper flow phenomena (Betts et al., 1991). It has been found that the inter-wrapper sodium flow phenomenon, especially on the storage SA, has a large potential for heat removal, and thereby the SSA sodium temperature is reduced considerably.

Detailed three-dimensional calculations carried out for a typical 500-MWe reactor have shown that consideration of inter-wrapper flow reduces the maximum sodium temperature reached after a loss of off-site power event followed by SCRAM (Parthasarathy et al., 2011). Figure 12.15 depicts the maximum central subassembly sodium, blanket subassembly sodium, and storage subassembly sodium temperatures with and without consideration of inter-wrapper flow (IWF). It can be observed that the beneficial effect of IWF heat transfer on SA temperatures is a reduction of about 20–60 K. The major differences are in the blanket and storage subassembly sodium temperatures. It was found that the natural convection flow was about 2–5% of nominal flow in the fuel and blanket SA and 10% in the storage region. This shows that non-consideration of IWF is conservative, but consideration of IWF gives a realistic and accurate estimation.

12.7 ROLE OF SECONDARY THERMAL CAPACITY

Further, the studies showed that a delay in the initiation of safety-grade decay heat removal (Parthasarathy et al., 2012) and a decrease in AHX air inlet temperature do not change the primary temperatures significantly. The secondary sodium inventory plays an important role in reducing the SA sodium outlet temperatures by about 30–60 K. Hence, one can get higher temperatures if there is total unavailability of the secondary sodium system, which is an event of very low probability. In this direction the upcoming European sodium fast reactor (ESFR) has each IHX linked to a steam generator in a casing through one secondary sodium circuit. There are six IHX, six SG, and six secondary sodium loops (Jeremy et al., 2021). In this manner even if one secondary loop is unavailable, the rest of the secondary loops would be available for decay heat removal through SG casing.

ASSIGNMENT

1. What are the contributions to decay heat in a nuclear reactor? Why is it important for safety?
2. What is the need to consider the natural convection flow of sodium in carrying out decay heat removal studies?
3. Considering the simple circuit shown in Section 12.2, derive the equation for natural convection flow and compare with the equation of Pavel Zitek (2014), given in the same section. (Hint: Take an average friction factor between laminar flow [proportional to Re] and turbulent flow [proportional to Re-0.2]. Also see Agrawal A.K. et al., Prediction of Decay Heat Removal Capabilities for LMBRs. *Nuclear Engineering and Design* 66 [I 98 I], 437–446.)
4. List different decay heat removal systems in a loop-type and pool-type SFR under different events.
5. In what manner can the reliability of a safety-grade decay heat removal system be improved, considering redundancy and diversity aspects? Is a passive system (natural convection) more reliable than an active system with a pump?
6. What is the role of inter-wrapper flow in minimizing temperature peaks after reactor trip under station blackout conditions? Explain in detail the processes taking place.

REFERENCES

Asteigno, J.C. et al. (1981), Theoretical and experimental analysis of SPX-1 thermal hydraulics problems in natural convection, *Decay Heat Removal & Natural Convection in FBRs*, Ed. Agrawal, A.K. and Guppy, J.G., pp. 275–287, Hemisphere Publishing Corporation.
Baldev Raj, Chellapandi P. and Vasudeva Rao P.R., (2015), *Sodium Fast Reactors in Closed Fuel Cycle*, CRC Press, https://doi.org/10.1201/b18350
Betts C. et al. (1991), European studies on fast reactor core inter-wrapper flows, *Proc. Int. Conf. on Fast Reactors and Related Fuel Cycles*, Vol. 3, Page P1.15.

Gouriou A., et al. (1982), Dynamic behavior of super PHENIX reactor under unprotected transient. In: Proceedings LMFBR Safety Topical Meeting, Lyon.

Idelchek I.E. (1993), *Handbook of Hydraulic Resistance*, 3rd edition, CRC Press.

Jeremy Bittan, Clement Bore and Joel Guidez (2021), Preliminary assessment of decay heat removal systems in the ESFR-SMART design: The role of natural air convection around steam generators outer shells in accidental conditions, *J. Nucl. Eng. Rad. Sci.* October, Vol. 7, p. 041301; https://doi.org/10.1115/1.4048991

Joël Guidez and Gérard Prêle (2017), *Superphenix Technical and Scientific Achievements*, Atlantis Press.

Kasinathan N., Athmaligam S., Vaidyanathan G., Chetal S.C. and Bhoje S.B. (1993), *Evaluation of DHR Studies for Indian FBR Program, IAEA Specialists Meeting*, Oarai, Japan, February.

Morgan V.T. (1997), Heat transfer by natural convection from a horizontal isothermal circular cylinder in air, *Heat. Transf. Eng.*, Vol. 18,pp. 25–33; https://doi.org/10.1080/01457639708939887

Natesan K. et al. (2013), Thermal hydraulic investigations of an extended station blackoutevent in FBTR, *Nucl. Eng. Des.*, Vol. 265, [[. 244–253; https://doi.org/10.1016/j.nucengdes.2013.07.022

Padmakumar G. et al. (2013), SADHANA facility for simulation of natural convection in the SGDHR system of PFBR, *Prog. Nucl. Energy*, Vol. 66, pp. 99–107, http://dx.doi.org/10.1016/j.pnucene.2013.03.019

Parthasarathy U., Sundararajan T., Balaji C. and Velusamy K. (2011), Estimation of effect of inter-wrapper flow heat transfer on core temperatures using porous body model, *Proceedings of the 21st National and 10th ISHMT-ASME Heat and Mass Transfer Conference*, December 27–30, IIT Madras, India.

Parthasarathy U., Sundararajan T., Balaji C. and Velusamy K. (2012), Development of a porous body model for decay heat removal studies in a pool type sodium cooled fast reactor, *Int. J. Adv. Eng. Sci. Appl. Math.*, September, DOI: 10.1007/s12572-012-0073-z.

Pavel Zitek and Vaclav Valenta (2014), Solution of heat removal from nuclear reactors by natural convection, *EPJ Web Conf.*, Vol. 67, p. 02133.

Schlunder E.U. (1983). Heat exchanger design handbook. United States: n.p., pp. 2153–2154.

Srinivasan G., Suresh Kumar K.V., Rajendran B. and Ramalingam P.V. (2006), The fast breeder test reactor—design and operating experiences. *Nucl. Eng. Des.*, Vol. 236, pp. 796–811; https://doi.org/10.1016/j.nucengdes.2005.09.024

Vaidyanathan G., Kasinathan K. and Velusamy K. (2010), Dynamic model of fast breeder test reactor, *Ann. Nucl. Energy*, Vol. 37, pp. 450–462; http://dx.doi.org/10.1016/j.anucene.2010.01.013

Vinod et al. (2013), Experimental evaluation of sodium to air heat exchanger performance. *Annals of Nuclear Energy*, Vol. 58, pp. 6–11.

Vinod V., Suresh Kumar V.A., Noushad I.B., Ellappan T.R., Rajan K.K., Rajan M. and Vaidyanathan G., (2006), Performance assessment of sodium to air finned heat exchanger for FBR, *Proceedings of ICONE 14 International Conference on Nuclear Engineering*, July 17–20, Miami, Florida, USA, https://doi.org/10.1115/ICONE14-89118.

Chapter 13

Modeling of Large Sodium–Water Reaction

13.1 INTRODUCTION

In the steam generator, sodium and water/steam are separated by a single tube wall. Any small hole or crack in the tube will lead to a sodium–water reaction. The sodium–water reaction produces heat and other reaction products, specifically NaOH, Na_2O, and NaH. These products cause corrosion and erosion of nearby tubes, leading to further escalation of water/steam leak rates (Hori, 1980).

Depending on the leak rate and its effect, the sodium–water reaction is classified into four types:

- Micro leaks (<0.1 g/s): A micro leak often develops from an intergranular crack in a defective weld and a fatigue crack in a tube wall. The corrosive reaction products often remain in place and plug the leak. The leak may stay plugged for several days or weeks.
- Small leaks (1–10 g/s): A small leak often causes localized tube damage, a phenomenon called *self-wastage*. The leak grows by a combination of corrosion and erosion. A small leak can also cause damage to adjacent steam tubes, a phenomenon termed *impingement wastage*. A water jet from a small leak forms a turbulent flow under sodium "flame."
- Intermediate leaks (10–2,000 g/s): When the leak rate rises into the intermediate range, the reaction flame becomes large and affects many other tubes. The flame interacts with the flowing sodium and triggers a chaotic turbulent interaction region characterized by widely fluctuating temperatures.
- Large leaks (>2,000 g/s): A large leak caused by the guillotine rupture of a tube or tubes may cause a rapid increase in pressure in the SG due to a large amount of hydrogen gas generated from SWR and damage the components of the secondary cooling circuits.

In the plant, the sodium–water reaction is immediately identified within the small leak range itself in sodium and in argon hydrogen detectors, and

DOI: 10.1201/9781003283188-13

safety actions are automatically initiated to stop further reaction. These include simultaneous isolation of water and steam side (in ~3 s), isolation of sodium inlet and outlet (~10 s), and dumping of water/steam inside the steam generator into a dump tank. Nitrogen is then injected into the tube side. These actions stop further sodium–water reaction.

In the intermediate leak range, pressure switches are provided in the surge tank. When the surge tank cover gas pressure increases above the threshold value, automatic actions are initiated to terminate further sodium–water reaction. In the large leak range (> ~2 kg/s), the major effect is the increased pressure in the sodium system (Greene, 1972).

For a large-size leak event, a rupture disk as a safety grade and sodium water reaction pressure relief system are designed to mitigate the event and prevent the sequential secondary SWR. Sodium leak detectors provided downstream of the rupture discs initiate the safety actions to terminate further sodium–water reaction. To ensure that the pressure surge gets attenuated before reaching the IHX, two capacities with free level and argon above sodium are provided. One is the surge tank in the hot secondary sodium leg, and the other is the pump tank in the cold secondary leg.

From overall reactor safety considerations, one needs to make certain that the secondary system design is adequate to ensure that the pressures generated in a large sodium–water reaction are attenuated well before reaching the IHX and that the IHX tubes are designed to withstand that pressure. This pressure specification needs an overall simulation of the following:

1. Water/steam leak rate
2. Reaction site dynamics
3. Propagation of pressure in secondary sodium
4. Reaction product discharge circuit

This chapter is devoted to analysis of a large sodium–water reaction to estimate the pressures in the different parts of the secondary sodium circuit. Application to the design of PFBR are presented.

13.2 LEAK RATE

13.2.1 Water Leak Rate Model

Analytical models exist that predict the maximum two-phase flow of saturated steam/water through tubes. A critical velocity exists at which the flow chokes in the minimum area, limiting the mass flow rate through the ruptured ends. Provided the external pressure is less than some specific limit, the mass flow rate is a function of the source pressure, length of the tube, and the area of rupture. Although the maximum flow rate of water is predictable for a given set of conditions, the manner of reaching this maximum flow

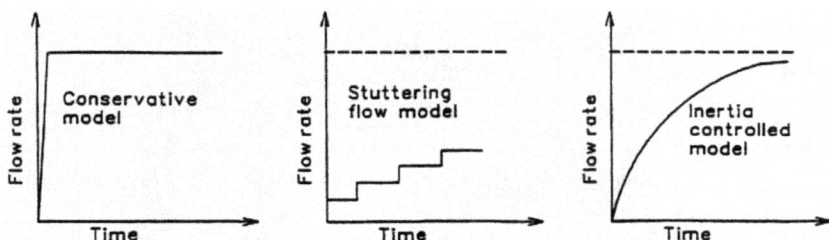

Figure 13.1 Water leak rate models.

rate is not. Three simple models for predicting the flow rates up to critical velocities are shown in Figure 13.1.

1. Conservative model: An ultraconservative estimate of the quantity of water injected over the initial period would be obtained by assuming the choked flow exists instantaneously.
2. Stuttering flow model: When the tube fails, the water leaps at some velocity through the rupture. This velocity is maintained while an expansion wave travels along the tube, gets reflected from water header, and approximately doubles the rupture exit velocity every time the reflected wave returns to the rupture site.
3. Inertia-controlled model: This model assumes the fluid to be incompressible and considers forces acting on an element of water in the tube. Water leaks from both ruptured ends and accelerates up to a critical velocity given by Burnell's relation (Tong, 1965).

The velocity of sound in water is 1,200 m/s. So, the transit time for the expansion wave to return to the rupture site has a minimum of 0.833 ms/m run of the tube. In two-phase mixtures, sonic velocity reduces to a very low value. This would give even longer transit times, and the stuttering flow model would predict very low flow rates for a long period following the rupture.

The first model, assuming choked flow exists immediately following the rupture, is ultraconservative and unrealistic in view of the expected transit times for the waves in the tube. The inertia-controlled flow model is a reasonable compromise between the other two models and has been adopted.

Balance of inertia forces with pressure forces, body forces, and frictional resistance describes the manner in which the water flow in upstream and downstream portions of the ruptured tube is accelerated up to a choked flow limit.

$$\frac{dq_u}{dt} = g_c A / L_u \left[(PH_u - PE_u) + \frac{g}{g_c} \rho \delta Z_u - k q_u^2 \right]$$

$$\frac{dq_d}{dt} = g_c A / L_d \left[\left(PH_d - PE_d \right) + \frac{g}{g_c} \rho \delta Z_d - k q_d^2 \right]$$

where

q_u and q_d are, respectively, the water flows upstream and downstream of
the ruptured tube,

PHu and PHd are, respectively, the upstream downstream water/steam
heard pressure,

PEu and PEd are, respectively, the end upstream/downstream portions of
the ruptured tube.

Flashing of the water occurs at the broken ends if the bubble pressure is less
than the saturated pressure of water.

Burnell's relation (Tong, 1965) gives the choked flow rate through the
broken ends,

$$q = A \left[2 g_c \rho \left\{ PE - \left(1 - C \right) p_{sat} \right\} \right]^{0.5}$$

$$C = 0.284 \sigma \left(T_w \right) / \sigma \left(196 \right)$$

13.2.2 Steam Leak Rate Model

Following a rupture in a superheater portion, steam flows toward both broken ends in the upstream and downstream portions of the ruptured tube.
Frictional adiabatic flow of steam has been assumed, since, during a period
of few hundred milliseconds, practically no heat gets transferred to steam.
Moreover, steam being lighter than water, the inertia effects caused by temporal acceleration are negligible in comparison to the inertia effects due to
spatial acceleration, pressure, and the frictional forces.

Above are the conditions required for a well-known "Fano flow" in ducts
(Shapiro, 1953). Utilizing these Fano flow relations and techniques available
for polynomial fitting, the steam flow rate,

$$\frac{q - q_{max}}{q_{max}} = \Sigma C_n \left[\frac{\phi - \phi_{cr}}{1 - \phi_{cr}} \right]^n$$

is obtained as a function of the pressure ratio, the critical pressure ratio, and
the maximum choked flow, in the form of a fourth-order polynomial. The
reaction site equations, characteristic equations for sodium flow adjacent to
the reaction site, and the water/steam leak rate equations are solved together
using the Runge-Kutta method.

13.3 REACTION SITE DYNAMICS

The reaction site dynamics modeling is based on the following features:

(1) The chemical rate of the sodium–water reaction is much higher than the rate of steam/water influx.

(2) The reaction between sodium and water is represented by the following general reaction equation:

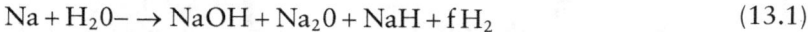

$$Na + H_2O- \rightarrow NaOH + Na_2O + NaH + fH_2 \tag{13.1}$$

where f is the ratio of moles of hydrogen produced to the moles of water reacted. This ratio and also the temperature at which hydrogen is generated are based on data used by different countries derived from their experimental programs (IAEA/IWGFR, 1983). In the present study the maximum value of 0.7 for f and 1,660 K for hydrogen temperature has been used.

(3) Hydrogen is taken to behave like an ideal gas, and its expansion at the reaction site is assumed to be adiabatic. This assumption is valid, as the expansion process is very fast.

(4) The bubble growth is analyzed using a combined model: (a) spherical bubble growth in an incompressible liquid in the vicinity of the reaction site, and (b) a one-dimensional axial motion (columnar) in a compressible liquid away from the reaction site (Figure 13.2). Such a model seems to be a realistic approximation because in the time span of the initial pressure spike the pressure wave bounces many times between the bubble and the wall and can be reasonably treated as growth in an incompressible liquid at the reaction site.

Based on the First Law of Thermodynamics, for a flow process (Rogers and Mayhew, 1967) we can write:

$$dQ = dH + dW \tag{13.2}$$

where: H = gas enthalpy = $C_p M_H T$, dW = work done by gas = $p\, dV$, dQ = heat of reaction, and M_H = molecular weight of hydrogen.

The heat of reaction brought into the reaction site by the hydrogen gas is specified by the temperature of generation (1,660 K) based on experimental observation as indicated earlier.

Hence,

$$dQ = C_p T_0\, dm_H,$$
$$dQ = C_p T_0 m_H \text{ where } m_H \text{ is mass of hydrogen}$$

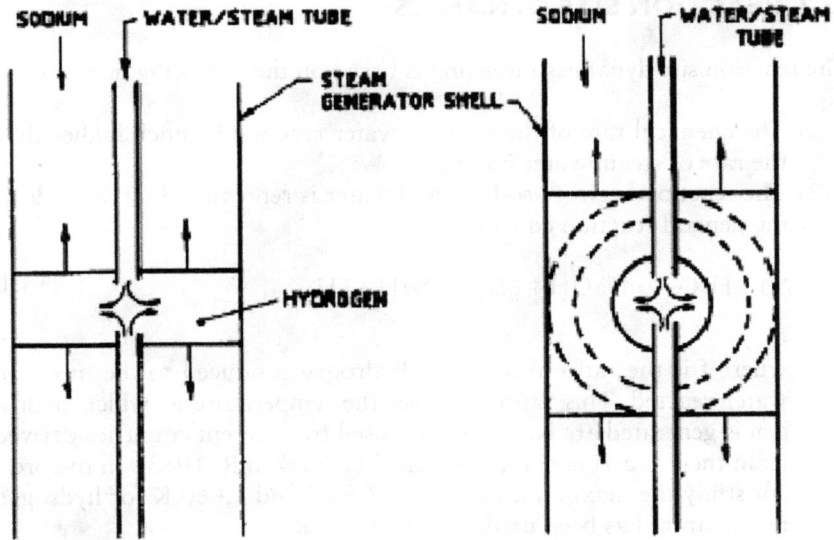

Figure 13.2 Reaction site model.

(*Source:* Selvaraj, P. et al., Large Leak Sodium-Water Reaction Analysis of an LMFBR Steam Generator Using a Variable Temperature Spherical Bubble Model. *Nucl. Eng. Des.* 123, 1990.)

The equation of state for hydrogen is

$$pV = \frac{m_H}{M_H}RT = \frac{fm}{M}RT \tag{13.3}$$

Here (m_H/M_H) for hydrogen is replaced by $(f.m/\underline{M})$ for water as per the general reaction Equation (13.1).

Substituting the value of H, dW, and dQ into (13.2) and eliminating p using Equation (13.3), we get

$$\frac{dT}{dt} = \frac{T_0 - T}{m}\frac{dm}{dt} - \frac{XT}{V}\frac{dV}{dt} \tag{13.4}$$

where $X = R/(C_p\,M_H)$. Differentiating Equation (13.3) and rearranging, we get

$$\frac{dp}{dt} = \frac{1}{V}\left[\frac{RT_0}{M}f\frac{dm}{dt} - (1+X)p\frac{dV}{dt}\right] \tag{13.5}$$

Here dV/dt is the bubble growth rate and dm/dt is the water leak rate and is given as input. Equations (13.4) and (13.5) are common for both first and second stages of expansion.

13.3.1 Spherical Bubble Model

The volume V of the bubble is given by $V = 4/3\pi r^3$ where r is the radius of the bubble. The bubble growth rate is

$$\frac{dV}{dt} = 4\pi r^2 u_r \tag{13.6}$$

where

$$u_r = \frac{dr}{dt} \tag{13.7}$$

Using Rayleigh's equation (Collier, 1981),

$$\frac{du_r}{dt} = \frac{p - p_1}{\rho r} - 1.5 u_r^2 / r \tag{13.8}$$

where P_1 is the pressure of sodium at the bubble–sodium boundary.

In view of the assumption that the bubble expands in an incompressible fluid at the reaction site, the same pressure will prevail throughout the sodium cross-section. The axial velocity of the liquid u_a is given by

$$\frac{du_a}{dt} = \frac{1}{\rho C} \frac{dp_1}{dt} \tag{13.9}$$

The axial and radial velocities are related using the volume balance

$$2A_c u_a = 4\pi r^2 u_r \tag{13.10}$$

where A_c is the cross-sectional area of the shell.

Ua and Ur are, respectively, the axial and radial velocities of sodium.

13.3.2 Columnar Bubble Model

In this case the volume V of the bubble is given by

$$V = 2A_c h \tag{13.11}$$

where h is the distance of the bubble–sodium interface from the leak point. The bubble growth rate is

$$\frac{dV}{dt} = 2A_c u_a \tag{13.12}$$

where

$$u = \frac{dh}{dt} = \frac{p - p_0}{\rho C} \tag{13.13}$$

where C is the sonic speed in liquid sodium.

13.3.3 Solution Technique

The leak rate is given as an input for the calculation. With this, there are seven independent variables, p, V, T, r, u_r, u_a, P_1, and seven Equations (13.4–13.10) for the first stage of expansion; and five independent variables, p, V, T, h, u_a, and five Equations (13.4), (13.5), and (13.11)–(13.13) for the second stage of expansion. These equations are solved using the Runge-Kutta method. The changeover from the first stage to the second stage occurs when the bubble diameter reaches the shell diameter. A computer code SWEPT has been developed incorporating these formulations.

13.3.4 Validation of Reaction Site Model

An experiment was conducted to study the effects of a large leak sodium–water reaction due to a guillotine failure of a single tube at the Brasimone sodium–water reaction test facility (Test BA12) jointly by France and Italy (Biscarel, 1982). The effects on the pressure and temperature variations and tube bundle deformations were studied. A small-scale model of the Super PHENIX steam generator helical coil type (shell diameter 1 m and height 2 m) was used in the experiment. The test was carried out by injection of subcooled water at 18 MPa and 573 K into static sodium at 0.13 MPa and 573 K. The water leak rate variation with time is presented in Figure 13.3. With this leak rate as a function of time, the model was used to calculate the reaction site pressure, and the same with legend SV is shown in Figure 13.4 and compared with experimental data. The reaction site pressure reaches a maximum value of 1.58 MPa at 3 ms (Figure 13.4).

The initial pressures calculated by SWEPT code are slightly higher than the experimental values. This is probably due to the discrepancy in the data on the leak rate of water. The leak rate reported (Biscarel, 1982) is itself a calculated one based on the sodium pressure and opening of the ruptured tube.

Further studies were carried out to understand the implications of different assumptions. The model studies were done assuming constant temperature. So, the effect of varying temperature on the pressure at the reaction site were explored. Studies by Hori (1980) had indicated that the assumption of a spherical model gives higher initial pressure than the columnar model. The results of the studies are also presented in Figure 13.4. The spherical bubble

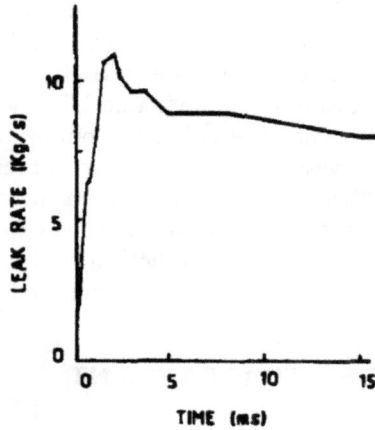

Figure 13.3 Water leak rate.

Source: Selvaraj, P. et al., Large Leak Sodium-Water Reaction Analysis of an LMFBR Steam Generator Using a Variable Temperature Spherical Bubble Model. *Nucl. Eng. Des.* 123, 1990.

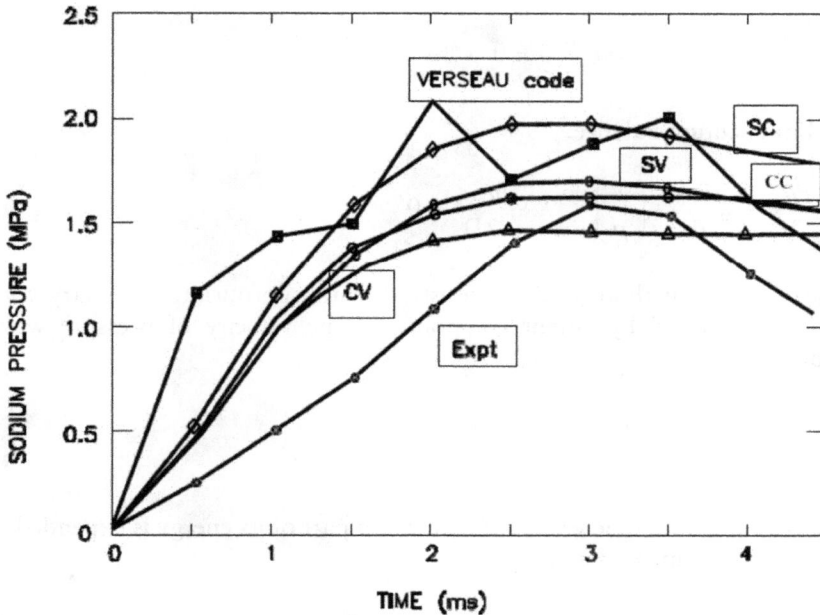

Figure 13.4 Comparison of different models, experiment and VERSEAU code.

Source: Selvaraj, P. et al., Large Leak Sodium-Water Reaction Analysis of an LMFBR Steam Generator Using a Variable Temperature Spherical Bubble Model. *Nucl. Eng. Des.* 123, 1990.

variable temperature (SV) model gives the peak pressure as 1.71 MPa at 2.83 ms, which is 8% higher than the experimental value. The spherical bubble model with constant temperature (SC) gives the highest value. The predicted maximum pressure is 1.98 MPa at 2.77 ms. This is 25% higher than the experimental value. The columnar bubble variable temperature (CV) model gives the lowest value: the maximum pressure is 1.46 MPa at 2.72 ms, which is 7.5% lower than the experimental value. The columnar bubble model with constant temperature (CC) gives 1.62 MPa at 3.73 ms (it is just 2.5% higher). It is clear that the spherical bubble variable temperature model gives reasonably conservative values compared to the experiment.

13.4 SODIUM SIDE SYSTEM TRANSIENT

Pressure waves emanating from sodium–hydrogen interfaces of the bubble get transmitted throughout the secondary circuit and interact with its components like sodium headers on top and bottom of SG, surge tank, pump tank, etc. This phenomenon of pressure wave propagation is described mathematically (Streeter and Wylie, 1979) by the simultaneous solution of one-dimensional transient equations of mass balance,

$$\frac{\partial \rho_{na}}{\partial t} + u_a \frac{\partial \rho_{na}}{\partial x} + \rho a^2 \frac{\partial u_a}{\partial x} = 0$$

and momentum balance,

$$\frac{\partial u_a}{\partial x} + u_a \frac{\partial u_a}{\partial x} + \left(\frac{1}{\rho}\right) \frac{\partial \rho_{na}}{\partial x} + \frac{f u_a u_a}{2D} = 0$$

utilizing the "method of characteristics" with appropriate boundary conditions described by circuit components. The velocity of pressure wave propagation,

$$a = C / \left(1 + \left(\frac{KD}{Ee}\right)\right)^{0.5}$$

is lower than the sonic velocity C because a part of its energy is expended in straining the pipe material.

13.5 DISCHARGE CIRCUIT SYSTEM TRANSIENT

Until the rupture disc ruptures, no calculation is carried out for the discharge circuit. When the rupture disc ruptures, initially only sodium flows through the rupture disc and the discharge circuit until the sodium–hydrogen interface reaches the rupture disc junction. Up to this point, calculations are carried out as described in Section 13.4. Once the interface reaches the rupture disc junction, thereafter two-phase sodium-hydrogen flow through the rupture disc and the discharge circuit is carried out. This is based on the homogeneous flow model in which the fluid is composed of an incompressible liquid and a compressible gas. No phase changes occur in the two components of the fluid. Depending on the available pressure drop, calculations are carried out either for critical two-phase flow or noncritical two-phase flow.

13.6 ANALYSIS OF PRESSURE TRANSIENTS FOR PFBR

The PFBR secondary sodium system consists of two identical independent loops. Each loop consists of a pump tank in which the secondary sodium pump is located, two IHX, a surge tank, and four SGs (Figure 13.5). SG is once-through integrated. Straight tubes with expansion bends in the bottom are used in the SGs. Sodium enters through a single inlet node and flows upward in the annular region, then flows down through the top inlet plenum. The tubes are placed in a triangular pitch. After flowing downward on the outside of the tubes, sodium exits through the bottom outlet plenum

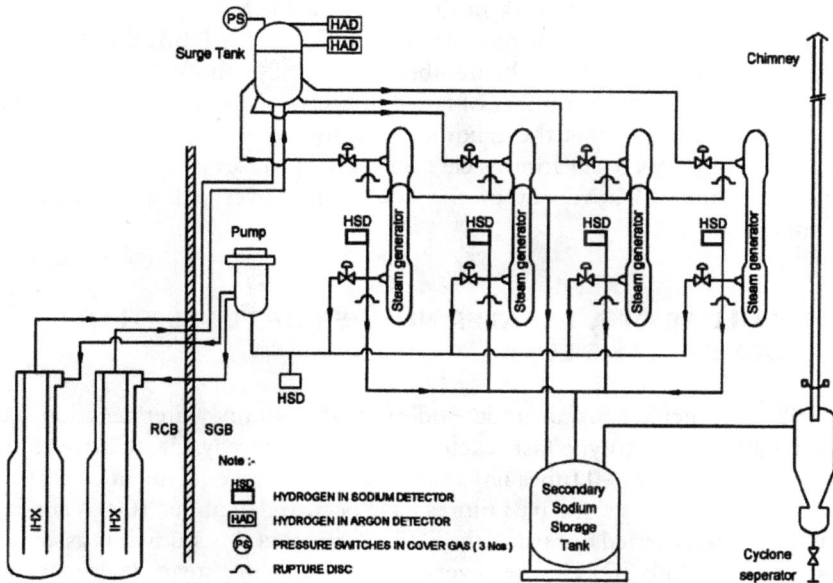

Figure 13.5 Secondary sodium circuit of PFBR.

Source: V. Sumathi, S. et al., Implications of Large-Scale Sodium Water Reactions in an LMFBR, *Nuclear Engineering and Design* 337, 2018.

and single outlet nozzle. Feed water enters the tube side at the bottom, flows upward in a counter-flow direction to the descending sodium. Each SG is provided with isolation valves on water and sodium sides.

For designing the secondary sodium circuit components, an upper bound of leak rate called *design basis leak* needs to be defined. Overheating of tubes occurs at the sodium–hydrogen interface. If the interface is stationary or moves very slowly, there will be enough time to cause continuous failure due to overheating. On the other hand, if the interface moves faster, there will be little or no time for overheating. In this way additional tube failures are prevented. The fast movement of the sodium–hydrogen interface is achieved by providing two rupture discs, one at the top and one at the bottom of the steam generator. The design basis leak is selected such that at that leak rate both the rupture discs should break almost simultaneously, leading to a fast movement of the sodium–hydrogen interface. After the pressure at the rupture disc reaches the set point, there is a delay in its response, which is based on actual tests. Experiments conducted for the FBTR reactor indicated a response time of 20 ms (Chetal et al., 1984).

A study has been carried out with the SWEPT computer code to estimate the number of simultaneous DEG failures required to ensure rupture of both rupture discs as well as pushing sodium away from the reaction site (Selvaraj et al., 1996). The leak is assumed to occur in the SG that is the farthest one from pump tank. For the leak at the top of the SG, the number of DEG failures required is 3. For the leak in the middle of the SG, the value is 4. For the leak at the bottom of the SG, the value is 5. For the DBA, the leak flow rate is more important than the number of tubes. For the leak at the top of the SG, even though the number of failed tubes is only 3, the leak rate is high at 13 kg/s. Because of this, the maximum pressures for the secondary sodium circuit components occur for this case only. For the DBA it takes only 5 s for all the sodium to move out of the SG. Hence, overheating failures are minimized.

13.7 FAILURE OF A GREATER NUMBER OF TUBES THAN DESIGN BASIS LEAK

In 1987, an incident of an under-sodium-leak sodium–water reaction has occurred in the prototype fast reactor (PFR) at Dounreay, UK (Currie et al., 1990). In this event, 40 tubes have failed, each one an equivalent to a DEG failure. Thirty-nine secondary failures have occurred in about 10 s. A similar study has been carried out using the SWEPT code. In this study it is assumed that one tube fails (DEG) after every 0.1 s. The water/steam leak rate per tube is 4.35 kg/s. After every 0.1 s, the total leak rate increases by 0.1 s, reaching a maximum value of 435 kg/s at 10 s. For the design basis leak case, the water/steam leak rate is 13 kg/s (three tubes fail) and remains constant. Figure 13.6 shows the pressure rise at the reaction site. It is obvious

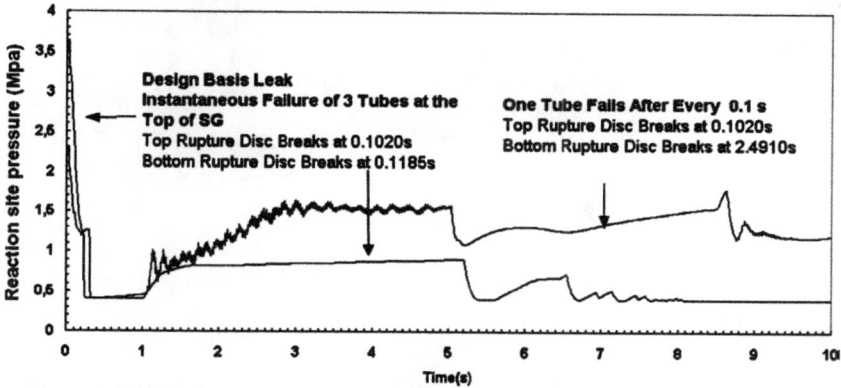

Figure 13.6 Multiple tube rupture studies.

Source: V. Sumathi, S. et Al., Implications of Large-Scale Sodium Water Reactions in an LMFBR. *Nuclear Engineering and Design* 337, 2018.

that the pressure rise at the reaction site is higher for the design basis leak. For the other case, even though there is a failure of 100 tubes in 10 s, the maximum pressure occurred is lower than that of design basis leak. This is mainly because the maximum pressure has occurred due to one tube failure only. Immediately afterwards, the maximum pressure reduces due to a rarefaction wave from the surge tank. Failure of the second tube coincides with the breaking of the top rupture disc. Once the top rupture disc ruptures, it opens a permanent relief path through the storage tank. The system now becomes open. Hence, any failure of further tubes could not increase the pressure above the initial peak value.

NOMENCLATURE

A_C area of cross-section of the steam generator
C   sonic speed in liquid sodium
C_p specific heat for hydrogen
h   distance of the bubble–sodium interface from the leak point
m   mass of water reacted
m_H mass of hydrogen produced
M   molecular weight of water
M_H molecular weight of hydrogen
p   bubble pressure
P_0 initial sodium pressure
P_1 pressure of sodium near the wall
R   universal gas constant
r   bubble radius

t   time
T   temperature of the bubble
T_o temperature at which hydrogen is generated
u_r radial velocity of sodium
u_a axial velocity of sodium
V   volume of bubble
X   constant = $R/(C_p M_H)$
ρ density of sodium

ASSIGNMENT

1. What causes a sodium–water reaction in the steam generator of SFR? Can it be avoided suitably by design?
2. What are the implications of a large sodium–water reaction?
3. What measures are taken in the system design of SFRs to minimize the pressure surge transmission to the primary sodium system? If the sodium–water reaction products enter the primary sodium and flow into the reactor, what sort of consequences can we expect?
4. Study the literature and compile a history of the sodium–water reaction and failures in different SFRs.

REFERENCES

Biscarel J. (1982), Pressure wave effects and SG tube bundle deformations following guillotine failure of a tube. In: Proc. Of LMIFBR tropical meeting, Lyon. Vol. III, p. 334.

Chetal S.C., Raju C., Anandkumar V., Seetharaman V. and Rajan K.K. (1984), Development of rupture discs for the FBTR, Liquid metal engineering and technology. BNES, London, pp. 305–308.

Collier J.G. (1981), *Convective Boiling and Condensation*, Mc Graw Hill Co., New York, pp. 118–120.

Currie R., Linekar G.A.B. and Edge D.M. (1990), The under-sodium leak in the PFR superheater 2 in February 1987. In: Proceedings of the Specialists' Meeting on Steam Generator Failure and Failure Propagation Experience, Aix-en-provence, France, IWGFR/78, pp. 107–132.

Greene D.A. (1972), Steam generator vessel pressures resulting from a sodium-water reaction: a computer analysis with the Swear code. *IAEA/INIS*, Vol. 3, pp. 218–231.

Hori M. (1980), Sodium/water reactions in steam generators of liquid metal fast breeder reactors. *Int. At. Energy Agency IAEA*, Vol. 12, pp. 707–778.

IAEA (1983), Proc. specialists' meeting on theoretical and experimental work on LMFBR steam generator integrity and reliability with a particular reference to leak development and detection, Hague, IAEA IWGFR/50.

Rogers G.F.C. and Mayhew Y.R. (1967), *Engineering Thermodynamics*, Longmans Green and Co. Ltd., London, pp. 32–35.

Selvaraj P., Seetharamu K.N. and Vaidyanathan G. (1990), Large leak sodium-water reaction analysis of an LMFBR steam generator using a variable temperature spherical bubble model. *Nucl. Eng. Des.* Vol. 123, pp. 87–90.

Selvaraj P., Vaidyanathan G. and Chetal S.C. (1996), Review of design basis accident for large leak sodium-water reaction for PFBR, *International Conference on Nuclear Engineering*, Volume 1 – Part B, ASME.

Shapiro A.M. (1953), *The Dynamics and Thermodynamics of Compressible Fluid Flow*, Ronald Press Company, New York.

Streeter V.L. and Wylie E.B. (1979), *Fluid Mechanics*. Mc-Graw Hill Book Co., New York.

Sumathi V., Jalaldeen S, Selvaraj P. and Murugan S. (2018), Implications of large-scale sodium water reactions in an LMFBR. *Nucl. Eng. Des.* Vol. 337, pp. 364–377. doi:10.1016/j.nucengdes.2018.07.013

Tong L.S. (1965), *Boiling Heat Transfer and Two-phase Flow*, John Wiley, and Sons.

Appendix A
Brief Description of Codes

The following codes were part of the IAEA coordinated program on inter-comparison of computer codes to analyze the PHENIX end-of-life tests (IAEA, 2013) and EBR II shutdown heat removal tests (IAEA, 2017)

A.1 SAS4A/SASSYS-1 CODE (ANL)

The SAS4A/SASSYS-1 code was developed at Argonne to perform safety analysis for liquid-metal-cooled reactors. SAS4A/SASSYS-1 contains extensive modeling capabilities that represent several hundred person-years of code development effort supported by experimental validation (Dunn et al., 1985).

These capabilities include:

Multiple channel and subchannel modelling of core thermal-hydraulics.

Point kinetics and spatial kinetics capabilities including decay heat and reactivity feedback models for fuel Doppler; fuel, cladding, and coolant density variations; coolant voiding; core radial expansion; control-rod driveline expansion; and primary vessel expansion.

Detailed mechanistic models for oxide fuel and cladding that characterize porosity migration, grain growth, fission gas release, fuel cracking with crack healing, fission-gas-induced swelling, irradiation-induced steel swelling, gas plenum pressurization, fuel-cladding gap conductance changes, fuel and cladding mechanical behaviour, thermal expansion, and cladding failure.

Detailed models of metallic fuel cladding transient behaviour, metal fuel pre-transient thermophysical properties characterization, and pre-failure transient behaviour models for fuel element mechanics, central cavity formation, extrusion, fission-gas-induced swelling, plastic flow, fuel-cladding eutectic formation, and fuel element failure detection.

Two-phase coolant thermal hydraulics model to characterize low-pressure sodium boiling with the ability to track the formation and collapse of multiple bubbles and the ejection of liquid slugs from coolant channels.

Intra-pin oxide fuel melting and relocation; cladding failure; molten cladding dynamics including melting, relocation, and freezing; fuel-coolant interactions in flooded channels including fission gas release, cladding perforation, molten fuel flow, and fuel freezing and plating; and fuel, fission gas, cladding, and coolant vapour dynamics in voided coolant channels.

Primary and intermediate loop reactor coolant systems models for compressible volumes (with or without cover gas), pipes, intermediate heat exchangers, centrifugal pumps, electromagnetic pumps, valves, bypass channels, annular flow elements, reactor vessel auxiliary cooling systems (RVACS), air-dump heat exchangers, and steam generators.

Balance of plant thermal hydraulics modelling capabilities including component models for deaerators, steam drums, condensers, reheaters, turbines, and several other components.

Reactor control system models that are driven by user-defined mathematical operators controlled by simulation variables.

SAS4A/SASSYS-1 version 3 has been exported to domestic industrial partners and to research organizations in foreign countries. The SAS4A/SASSYS-1 code package continues to undergo development in response to advanced fast reactor simulation needs.

A.2 CATHARE CODE AND SPECIFIC MODEL FOR SFR CALCULATIONS (CEA AND IRSN)

The system code CATHARE has been developed and validated in collaboration between CEA, EDF (electricity supplier), IRSN (French safety authority), and AREVA-NP (plant manufacturer) for the French Pressurized Water Reactors. The CATHARE code is the reference code in France for the PWR safety analysis. It has also been used for other light water reactor concepts (VVER, BWR, RBMK) and for experimental reactors applications.

It is a 6-equations code for two-phase flows (2 equations of mass balance, 2 equations of energy balance and 2 equations of momentum balance). The main variables are the pressure, the liquid and gas enthalpies, the liquid and gas velocities, the void fraction (and possibly mass fraction of incondensable gases). The aim of the code is to represent mechanical non-equilibrium and thermal non-equilibrium, at all flow regimes and all heat transfer regimes within the range of design and safety analysis. The code uses an oriented-object structure permitting the representation of any kind of hydraulic circuit, from the analytical experimental facilities to the whole reactor power plant. There are five main modules, with specific correlations and closure laws (1D: axial; 0D: volume, THREED or 3D: PWR reactor vessel, BC: boundary condition and RG: double ended break). The code numerical scheme is fully implicit (1D and 0D) or semi-implicit (3D), with an implicit thermal coupling between the walls and the fluid. The non-linear system is solved by a Newton-Raphson iterative method.

In the Generation IV framework, the standard version of the code (CATHARE_2v2.5_2) has already integrated new developments in order to calculate gas cooled reactors and super critical light water reactors. The development of CATHARE code for sodium-cooled reactors has followed this trend (Geffraye et al., 2009).

The main developments in CATHARE code to address sodium-cooled reactor calculations deal with:

Sodium fluid properties (the thermodynamic and transport properties of liquid and vapor are issued from former CEA system codes);
Specific heat exchange correlations (Skupinski and Borishanski heat exchange correlations);
Specific friction loss coefficient in the fuel pins zone (Pontier's law);
Update of the neutronics points kinetics model considering neutronics feedback effects due to diagrid expansion, wrapper axial and radial expansion, and fuel clad axial and radial expansion.

The CATHARE code is ready for use for sodium-cooled reactor calculations and new concept evaluations. CATHARE is the thermal hydraulics system code used in the frame of the ASTRID project, the Generation IV French prototype project.

A.3 STAR-CD AND DYANA-P SYSTEM CODES (IGCAR)

STAR-CD is a commercial CFD code that solves the governing differential equations of flow physics by numerical means on a computational mesh (CD-ADAPCO, 2005). It has the capability of solving steady, transient, laminar, turbulent, compressible, and incompressible flow phenomena along with heat transfer (convection, conduction, and radiation) even in a porous medium. It has an in-built pre-processor and post-processor known as PROSTAR. It has the basic mesh generation capability. Complex mesh can be imported from any standard mesh-generating tools. User-defined program modules can be added to the code to modify the material properties as well as pressure drop and heat transfer characteristics dynamically during transient. The code has been validated extensively against benchmark data.

DYANA-P is a system dynamics code developed by IGCAR for performing plant dynamics studies for pool-type fast breeder reactor systems (Natesan et al., 2011). This is the design tool used for the safety analysis of prototype fast breeder reactor (PFBR). DYANA-P has models for reactor core, primary sodium circuit, secondary circuit, sodium pumps, heat exchangers, and sodium pools. The steam water system is not modeled in the code. However, a steam generator is modeled with appropriate boundary conditions for the water side.

Mathematical models in the DYANA-P code are based on those used for most of the fast reactors. Thermal models are based on heat balance between various sections exchanging the heat such as fuel and sodium through the clad in SA, primary sodium, and secondary sodium through the tube wall in IHX, sodium and ambient air through the pipe wall and insulation in sodium piping, secondary sodium and water through the tube wall in steam generators (SG), etc. The hydraulic model is based on momentum balance between various flow segments in the primary and secondary sodium circuits. The torque balance is adopted for the modeling of the pump with the characteristics derived from generalized homologous characteristics. Fluid levels in the tanks are modeled through the dynamic mass balance. The neutronic model for the core is based on a PK approximation. The transient solution is obtained via a prompt jump approximation. Detailed models are also incorporated for the calculation of various reactivity feedback effects due to radial expansion of grid plate where SA are supported, control rod expansion, volumetric expansion of sodium inside the fuel and blanket portion of core, axial clad steel expansion, axial fuel expansion, and the Doppler effect due to changes in fuel temperature.

Major assumptions made in the mathematical models of the various reactor and heat transport components are:

Liquid sodium flow is assumed to be incompressible and single phase throughout.

The flow of sodium is treated as one-dimensional through the pipelines, fuel rod bundles, heat exchanger tubes, etc.

In all the places where mixing and recirculating flow patterns exist, a perfect mixing assumption is made for the thermal model, and incoming kinetic energy head due to flow is assumed to be fully converted into static pressure head for hydraulic calculation.

Axial conduction heat transfer in coolant pipe material, heat exchanger tubes, fuel pin, and clad walls are neglected in comparison to radial heat transfer.

Initial steady-state conditions are calculated in the code through the solution of steady-state balance equations. For the transient solution, numerical integration of hydraulic models is obtained by utilizing a standard ordinary differential equation solver based on the Hamming's predictor-corrector method. Semi-implicit formulation has been adopted for the integration of thermal balance equations.

A.4 GRIF CODE

GRIF is a Russian computational tool for single-phase thermal hydraulics analysis of the transients in the reactor as a whole and in its parts (Chvetsov and Volkov, 1998) The code can be characterized as a system code with extended capabilities for the simulation of spatial distributions of most important parameters in reactor bulk volumes and in structural elements. One of important objectives of the GRIF code is the simulation of long-term processes in the reactor including the secondary and auxiliary circuits (decay heat removal problem etc.).

The code includes the following modules:

3D thermal hydraulic model for calculation of sodium velocity, pressure, and temperature in the primary circuit.

3D model for simulation of inter-wrapper sodium thermal hydraulics.

Primary pump model (analytical correlation).

Module "Wrapper" for calculation of temperature distributions in the SA wrappers.

Module "IHX" for simulation of flow and temperature in the IHXs.

Module "DHX" for simulation of flow and temperature in the DHXs (module is not activated in PHENIX test simulation).

Module "PIN" consists of the set of option of 1D-3D models that optionally can be used for calculation of temperature in fuel pins, absorber pins, shielding elements simulation, etc.

Module "REACTIVITY," analytical correlations with recalculated reactivity coefficients (module is not activated in PHENIX test simulation).

Module "KINETICS" point kinetics with six groups of delayed neutrons (module is not activated in the PHENIX test simulation).

For simulation of heat and mass transfer in the reactor, the set of 3D heat and mass transfer equations in approximation by the model of viscous non-compressible liquid flowing in the porous body is solving in GRIF code. The non-isothermal conditions effect is considered using a business approximation.

Porosity of the coolant medium is varying spatially, and the porous medium resistance coefficients may depend on flow parameters. The thermal hydraulic properties—effective kinematics viscosity, coolant density, and specific heat capacity—are the functions of sodium temperature. Effective conductivity coefficients of porous medium can be different for different directions.

The similar set of equations is solved for simulation of sodium behavior in the inter-wrapper space, the only difference being that mass and heat flux sources in the equations have opposite signs. Mass source value is different from zero only on the outer circuit of subarea modeling reactor core, since two sodium flows (the main flow in the subassemblies and that in inter-wrapper space) are sewn together only on the boundaries. Heat transfer between these sodium flows occurs in the entire core volume.

The above two sets of equations have been solved numerically, and the solutions are sewn together explicitly by iteration method.

A.5 NETFLOW++ CODE (UNIVERSITY OF FUKUI)

The system analysis 1D code NETFLOW++ has been developed by Mochizuki (Mochizuki, 2007, 2010). The code can calculate single-phase and two-phase flows of water. One-dimensional flow with no compressibility is assumed in the single-phase flows of water or liquid sodium. Piping is divided into some segments called main links by main joints. The main link can be divided into smaller segments called *sublinks* by subjoints. When the subjoint is provided, one can change the diameter of piping between sublinks. The main joint can connect several main links, but a subjoint cannot. Other than these joints, two kinds of joints for setting pressure boundary conditions and flow rates are prepared. Simultaneous mixing is assumed at the joint.

The one-dimensional momentum and continuity equations are applied to the flow segment together with the energy equation. In the temperature calculation of piping and heat transfer tubes of a heat exchanger, the material is divided into inside and outside at a half thickness in order to have symmetrical equations for both sides. Temperature is calculated based on the heat transfer between fluid and pipe wall, and also pipe wall to environment or the flow outside the heat transfer tube. Heat transfer coefficient is given to the code as a function of the Reynolds number and the Prandtl number.

Several types of heat exchangers are modeled, e.g., a usual shell-and-tube type, a shell-and-tube type with boundary conditions on one side, an air cooler with finned heat transfer tubes, and steam generators. In the case of a steam generator, a heat transfer coefficient for a helically coiled heat transfer tube is prepared other than a straight type. When a pump is provided on the sublink, the pressure increase occurs. The pump characteristic is expressed as a H-Q curve, and the pressure head is approximated as a function of quadratic volumetric flow rate. In the pump characteristics evaluation for steady state and transients such as pump start-up and coast-down, the kinetic equation with pump efficiency is solved. In regard to a valve, two kinds of inputs are prepared. One is a timetable of local loss coefficients. The other one is a combination of a valve characteristic Cv as a function of throttling and a timetable of the valve throttling. A check valve characteristic is also modeled.

Models in the code are verified using test results of mockups of a boiling water reactor. Thermal hydraulics and neutronics of the coupled system of the core, heat transport systems, and the turbine system can be calculated by this code. The models in the code have been developed further in order to simulate some characteristics specific to liquid metal–cooled fast reactors. These are the model of inter-subassembly heat transfer that becomes obvious under the natural circulation condition, the heat transfer and heat transfer models for the air cooler, the heat transfer model for IHX, the model of the upper plenum, etc. The code has been validated using the sodium-cooled experimental fast reactor Joyo and the prototype fast breeder reactor Monju, i.e., the natural circulation test with Mark-II core of Joyo, the intentional scram test at Joyo with Mark-III core, the natural circulation test in Monju using pump heat input, and the turbine trip test at Monju. One of the characteristics of the code is fast running using PC. A one-week event of three-loop Monju can be calculated within 30 minutes. This code is used for the education of students at the University of Fukui.

A.6 MARS-LMR CODE (KAERI)

The MARS-LMR code has been developed by the KAERI for the design and analysis of a liquid metal–cooled system based on the MARS (multidimensional analysis for reactor safety), which has been widely used for the analysis of transients of water-cooled reactor systems in the Republic of Korea (Ha et al., 2007). MARS is a multi-D, transient, two-fluid, six-equation model for flow of a two-phase steam-water mixture that can contain noncondensable components in the steam phase and/or a soluble component in the water phase. The two-fluid equations are formulated in terms of volume and time-averaged parameters of the flow. The primary variables are the single pressure, two-phase internal energies, and the void fraction at the scalar control volume and two-phase velocities at the momentum control volume. The code employs a semi-implicit numerical scheme to solve a nonequilibrium and nonhomogeneous thermal hydraulic problem. The heat structure can be coupled either implicitly or explicitly with the fluid, and the code contains the point kinetic module.

The code continued to evolve by adding new features such as multidimensional module, CANDU specific models, HTGR specific models, and LMR models. In the Gen-IV framework, to extend the applicability of the code to the analysis of transients for a sodium-cooled fast reactor system, it is required to reinforce some models that represent the inherent characteristics of liquid sodium–cooled fast reactors. First, liquid sodium has quite different characteristics from water. Therefore, it is necessary to add several new models, such as a liquid metal coolant property table, wall heat transfer coefficients related to liquid metal, and the friction factor correlations associated with wire spacers of fuel rod were implemented to the code. The material properties tables are implemented using the equation of state (EOS) based on the semi-empirical soft-sphere model. Other dynamic properties such as viscosity, conductivity, and surface tension are provided by the relevant correlations. For a description of the pressure drop in a wire-wrapped rod bundle, the correlation by S.C. Cheng and N. E. Todreas has been implemented in the MARS-LMR code. The heat transfer by a liquid metal flow in nuclear fuel bundles is described by the modified Schad's correlation. For the liquid metal heat transfer in a heat exchanger tube, the correlation by Aoki is selected, and it is found that the Graber-Rieger model is best for a shell side heat transfer

A.7 TRACE CODE (PSI)

TRACE (Spore et al., 2001) is a thermal-hydraulics code developed by the U.S. Nuclear Regulatory Commission (NRC). Formerly called TRAC-M, TRAC/RELAP Advanced Computation Engine is the latest in a series of advanced, best-estimate reactor system codes. It combines the capabilities of the NRC's four main systems codes (TRAC-P, TRAC-B, RELAP5, and RAMONA) into a single modernized computational tool. TRACE has been originally designed to perform best-estimate analyses of loss-of-coolant accidents (LOCAs), operational transients, and other accident scenarios in PWRs and BWRs. Also, its versatility allows one to model a wide variety of thermal-hydraulics experiments in reduced-scale facilities. Models used include multidimensional two-phase flow, non-equilibrium thermodynamics, generalized heat transfer, reflood, level tracking, and reactor kinetics. The programming language is standard Fortran 90.

The partial differential equations that describe two-phase flow and heat transfer are solved using finite-volume numerical methods. The heat transfer equations are evaluated using a semi-implicit time-differencing technique. The fluid-dynamic equations in the spatial 1D, 2D, and 3D components use a multistep procedure (SETS numerics). The finite-difference equations for hydrodynamic phenomena form a system of coupled, nonlinear equations that are solved by the Newton-Raphson iteration method. The resulting linearized equations are solved by direct matrix inversion.

The modeling of a reactor system is based on a component approach. Each physical piece of equipment in a flow loop can be represented as some type of component, and each component can be further nodalized into some number of physical volumes (cells) over which the fluid, conduction, and kinetics equations are averaged. The number of reactor components in a problem and the manner in which they are coupled are arbitrary. The only limit on the problem size is the amount of computer memory. The hydraulic components available in TRACE for the reactor description include, among others, PIPEs (1D), VESSELs (3D), PLENUMs (0D), PUMPs, VALVEs, and TEEs. The fuel elements and heated walls in the reactor system can be modeled with HTSTRs (heat structures) that compute 2D conduction and surface-convection heat transfer in X-Y or R-Z geometries. The energy delivered to the fluid via the HTSTRs is specified in the POWER components. The power generation in the reactor core can be specified in different ways: either constant or via a time-dependent table, calculated from the point-reactor kinetics with reactivity feedbacks, or calculated from 3D kinetics (when TRACE is used in coupled mode with PARCS that provides the reactor power and power distribution at each time step). The boundary conditions in the hydraulic components are applied from the FILL and BREAK components, used to apply the desired coolant flow and pressure boundary conditions, respectively.

The TRACE code is currently being extended at PSI to sodium two-phase flow models in order to allow the simulation of a boiling event in SFRs. Extensive validation of the implemented model has been performed on the basis of past out-of-pile experiments (Chenu et al., 2009, 2011). The analysis of the PHENIX NC test constitutes a step toward further validation of the TRACE code for SFR transient simulation.

A.8 SAC-CFR (NCEPU)

The system analysis code for China fast reactor (SAC-CFR) (Lu et al., 2012; Sui et al., 2013) was developed for fast reactors to predict the plant response during operational and accidental transients. The main components in the primary loop, intermediate loop, and tertiary loop are modeled to calculate the response of neutron kinetics, thermal hydraulic, plant control, and the protection system. The response includes temperatures and mass flow rates in the core and the loop, fuel temperature, pump performance, heat exchanger performance, etc. The SAC-CFR code is suited for different kinds of transients ranging from normal operational conditions to upset conditions caused by such disturbances as loss of flow and loss of heat sink.

A.9 THACS (XJTU)

The 1D system analysis code THACS (transient thermal hydraulic code for analysis of sodium-cooled fast reactor) has been developed by Xi'an Jiao tong University (Ma et al., 2015; Yue et al., 2015). The code can calculate single-phase and two-phase flows of sodium. One-dimensional flow with non-compressibility is assumed in the single-phase flows of sodium. For the two-phase calculation, the multi-bubble model is used in the core module for sodium. A compressible water model is applied on the water side of steam generators. THACS uses an object-oriented structure permitting the representation of any kind of hydraulic circuit, from the analytical experimental facilities to the whole reactor power plant. The code capabilities are listed below:

(a) Multiple-channel core thermal hydraulic analysis
(b) Point-kinetic resolution covering decay heat and reactivity feedback models, including fuel Doppler, fuel and coolant density variations, core radial expansion, control rod driveline expansion, and coolant voiding
(c) Models of metallic fuel and MOX fuel thermo-physical properties
(d) Primary and intermediate loops of reactor coolant systems models such as pipes, intermediate heat exchangers, centrifugal pumps, hot pools and cold pools, pipe-nets, air-dump heat exchangers, steam generators, inter-wrapper flow, and reactor vessel auxiliary cooling systems.

REFERENCES

CD-ADAPCO, (2005), "*Star CD code Version 3.26*", Computational Dynamics Limited, New York, USA.

Chenu A, Mikityuk K. and Chawla, R., (2009), "TRACE simulation of sodium boiling in pin bundle experiments under loss-of-flow conditions", *Nucl. Eng. Des.*, Vol. 269, pp. 2417–2429.

Chenu A., Mikityuk K. and Chawla, R., (2011), "Pressure drop modeling and comparisons with experiments for single- and two-phase sodium flow", *Nucl. Eng. Des.*, Vol. 241, pp. 3898–3909.

Chvetsov I. and Volkov A., (1998), "3-D thermal hydraulic analysis of transient heat removal from fast reactor core using immersion coolers", *IAEA Technical Committee Meeting on Methods and Codes for Calculations of Thermal hydraulic Parameters for Fuel, Absorber Pins and Assemblies of LMBFR with Traditional and Burner Cores*, Obninsk, Russian Federation, 27–31 July 1998.

Dunn F.E., Prohammer F.G., Weber D. P. and Villim R.R., (1985), "The SASSYS-1LMFBR systems analysis code", *Proc. of Intl. Topl. Mtg. Fast Reactor Safety*, Knoxville, Tennessee, pp. 999–1008.

Geffraye G., Farvacque M., Kadri D., Lavialle G., Ruby A., Rameau B. and Antoni O., (2009), "CATHARE 2 V2.5_2: a single version for various applications", Proc. Of NURETH-13, Kanazawa City, Ishikawa Prefecture, Japan, 27 September–2 October 2009.

Ha K.S., Jeong H.Y., Chang W.P., Lee Y.B. and Jo C.H., (2007), "Development of MARS-LMR and steady state calculation for KALIMER-600", Technical Report No. KAERI/TR-3418/2007, Korea Atomic Energy Research Institute, Rep. of Korea.

IAEA-TECDOC-1703, (2013), "Benchmark analyses on the natural circulation test performed during the Phenix end-of-life experiments-final report of a coordinated research project 2008–2011", IAEA, Vienna.

IAEA-TECDOC-1819, (2017), "*Benchmark Analysis of EBR-II Shutdown Heat Removal Tests*", IAEA, Vienna.

Lu, D., Sui, D.T., Ren, L.X., Qian, H.T. and Tian, L., (2012), "Development of system analysis code for pool-type fast reactor under steady state operation", *Atom. Energy Sci. Technol.*, Vol. 46, No. 4, pp. 422–428.

Ma Z., Yue N., Zheng M., Hu B., Su G. and Qiu S., (2015), "Basic verification of THACS for sodium cooled fast reactor system analysis", *Ann. Nucl. Energy*, Vol. 76, pp. 1–11.

Mochizuki H., (2007), "Verification of NETFLOW code using plant data of sodium cooled reactor and facility", *Nucl. Eng. Des.*, Vol. 237, No. 1, pp. 87–93.

Mochizuki H., (2010), "Development of the plant dynamics analysis code NETFLOW++", *Nucl. Eng. Des.*, Vol. 240, pp. 577–587.

Natesan K., Kasinathan N., Velusamy K., Selvaraj P., Chellapandi P. and Chetal S.C., (2011), "Dynamic simulation of accidental closure of intermediate heat exchanger isolation valve in a pool type LMFBR", *Ann. Nucl. Energy*, Vol. 38, pp. 748–756.

Spore J.W., et al., (2001), "*TRAC M FORTRAN 90 (Version 3.0) Theory Manual, Division of Systems Analysis and Regulatory Effectiveness*", Office of Nuclear Regulatory Research, U.S. Nuclear Regulatory Commission, Washington DC.

Sui D., Lu D., Ren L. and Liu Y., (2013), "Development of three-dimensional hot pool model in a system analysis code for pool-type FBR", *Nucl. Eng. and Design*, Vol. 256, pp. 264–273.

Yue N., Ma Z., Cai R., Hu B., Su G.H. and Qiu S., (2015), "Thermal-hydraulic analysis of EBR-II shutdown heat removal tests SHRT-17 and SHRT-45R", *Prog. Nuc. Energy*, Vol. 85, pp. 682–693.

Appendix B
Primary Pump Discharge Pipe Break Modeling

B.1 INTRODUCTION

The primary sodium pipes connect the pumps to the core inlet plenum. The design of the piping is done under safety class 1 rules and can withstand a design basis earthquake. With high-quality manufacturing standards, breakage of a pipe is a very low-probability event. In loop-type reactors, all primary pipelines are provided with double envelopes and interspace coolant leak monitoring systems that permit leak detection before the break. Thus, the PCP rupture event can be placed in the beyond design basis event (BDBE) category. Such an arrangement is difficult to incorporate for pool-type reactors. Hence, due to the difficulties of leak detection and justification of leak before break (LBB), a double-ended guillotine rupture of the pipe is considered as a category 4 design basis event for pool-type reactors. The response of the plant protection system to the accident condition and details of transient flow through the core during the first few seconds after pipe break are important in determining the severity of the transient response. There is a need to analyze this event and ensure that the event is detected in time and reactor shutdown effected. The transient following the event is a sharp reduction in core flow and consequent rise in core temperatures. Analytical studies have been caried out by different countries for this event (IAEA/IWGFR, 2003). The primary hydraulics model used in the dynamic studies needs to be separately studied and linked to the rest of the plant dynamics model to analyze the consequences for the plant. In the following the mathematical modeling of the flow through the various segments of the primary sodium circuit along with the solution technique adopted for PFBR is briefly described.

B.2 MODEL EQUATIONS

PFBR has two primary pumps and four intermediate exchangers in the pool. From each pump two pipes emerge that join the grid plate (Figure B.1).

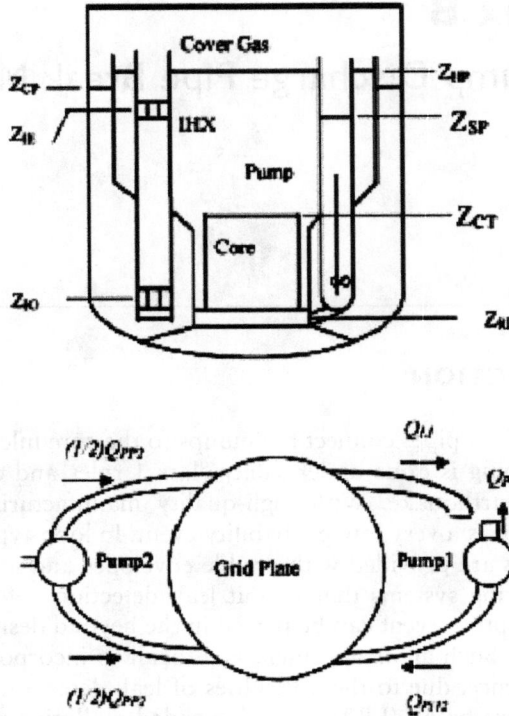

Figure B.1 Primary sodium circuit with rupture of one pump discharge pipe.

The specific model features are:

The incipient cavitation flow and the slope of the head versus flow line at the rated speed of the pump is available from the pump design data.

The effects of radial pressure distribution in the inlet plenum on the flow distribution among the fuel SA are neglected. This is due to relatively low pressure drop in the inlet plenum compared to that in the fuel SA.

Torque consumed at the impeller under cavitating condition is assumed to be equal to the value evaluated at the incipient cavitation point.

The pressure drop at the ruptured end of the discharge pipe is modeled by using a velocity head loss coefficient 1.

The hydraulic resistance network of the primary circuit is presented in Figure B.2.

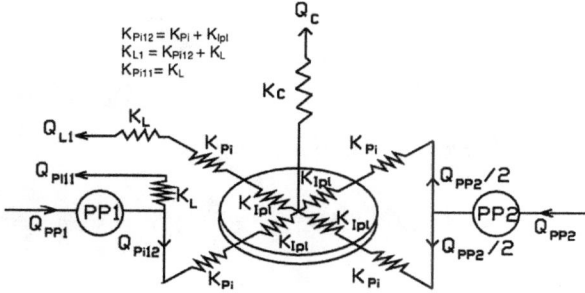

Figure B.2 Hydraulic resistance network of primary with one pipe rupture.

The governing equations of momentum balance between different segments are (Natesan et al., 1999):

- Between hot pool and IHX

$$A_I \frac{dQ_{Il}}{dt} = (Z_{HP} - Z_{IE})\rho_{HP}\,g - (Z_{CP} - Z_{IO})\rho_{CP}\,g - K_I Q_{Il}/Q_{Il}/ - g$$

$$\int_{Z_{IE}}^{Z_{IO}} \rho_l dZ.\ for\ l = 1,\ 2 \qquad\qquad (1\ and\ 2)$$

where Q_{LI} is the flow through IHX and Z_{HP}, Z_{CP}, Z_{IE}, Z_{IO} are the elevations of hot pool free sodium level, cold pool free sodium level, IHX primary inlet, and IHX primary outlet, respectively.

- Between pump 1 and discharge pipe junction

$$A_{PP1} \frac{dQ_{PP1}}{dt} = (Z_{CP} - Z_{IO})\rho_{CP}\,g - P_{J1} + D_{pp1} - K_{PPS}\,Q_{PP1}/Q_{PP1}/ \qquad (3)$$

- Between pump 1 discharge pipe junction and cold pool via broken pipe

$$A_{Pi11} \frac{dQ_{Pi11}}{dt} = P_{J1} - (Z_{CP} - Z_{IO})\rho_{CP}\,g - (K_{pi11} + K_L)Q_{Pi11}/Q_{Pi11}/ \qquad (4)$$

- Between pump 1 discharge pipe junction and grid plate

$$A_{Pi12} \frac{dQ_{Pi12}}{dt} = P_{J1} - \Delta P_{core} - (Z_{HP} - Z_{CT})\rho_{HP}\,g$$

$$- (K_{Pi12} + K_{Ipl})QPi12/Q_{Pi12}/ \qquad\qquad (5)$$

where $\Delta P_{core} = P_{CE} - P_{CT}$

- Between grid plate and broken pipe to cold pool

$$A_{L1}\frac{dQ^{L1}}{dt} = \Delta P_{core} + \left(Z_{HP} - Z_{CT}\right)\rho_{HP}\ g - \left(Z_{CP} - Z_{IO}\right)\rho_{CP}\ g \\ - \left(K_{L1} + K_L + K_{Ipl}\right)Q_{L1}/Q_{L1}\ / \tag{6}$$

- Between pump 2 and grid plate

$$A_{PP2}\frac{dQ_{PP2}}{dt} = \left(Z_{CP} - Z_{IO}\right)\rho_{CP}\ g - \left(Z_{HP} - Z_{CT}\right)\rho_{HP}\ g \\ - \Delta P_{core} - K_P\ Q_{PP2}/Q_{PP2}/ + D_{PP2} \tag{7}$$

Using the relation $\dfrac{dQ_{PP1}}{dt} = \dfrac{dQ_{Pi11}}{dt} + \dfrac{dQ_{Pi12}}{dt}$ (8)

and $\displaystyle\sum_{r=1}^{10} n_r \frac{dQ_{cr}}{dt} = \frac{dQ_{Pi12}}{dt} + \frac{dQ_{PP2}}{dt} - \frac{dQ_{L1}}{dt}$ (9)

a detailed expression for ΔP_{core} and P_{J1} can be obtained:

$$\rho_{HP}\ S_{HP}\frac{dZ_{HP}}{dt} = \left[\sum_{l=1}^{2}\left(Q_{PPl} - Q_{ll}\right)\right] - \left(Q_{L1} - Q_{Pi11}\right) \tag{10}$$

$$\rho_{CP}S_{CP}\frac{dZ_{CP}}{dt} = \left[\sum_{l=1}^{2}\left(Q_{ll} - Q_{PPl}\right)\right] + \left(Q_{L1} + Q_{Pi11}\right) \tag{11}$$

Energy balance on pump shaft can be represented by

$$I_{PP}\frac{d\omega_{PPl}}{dt} = \beta_{PDrl} - \beta_{PPl}, \quad for\ l = 1,2 \tag{12 and 13}$$

$$D_{PPl} = f1\left(Q_{PPl}, \omega_{PPl}\right), \quad for\ l = 1,2 \tag{14 and 15}$$

$$\beta_{PPl}\left(tot\right) = \beta_{PPl}\left(impeller\right) + \beta_{PPl}\left(fri\right), \quad for\ l = 1,2 \\ = f2\left(Q_{PPl}, \omega_{PPl}\right) + \beta_{PPl}\left(fri\right) \tag{16 and 17}$$

where f_1 and f_2 are taken from the homologous characteristic.

Considering 10 flow zones in the core, one can write the following equations for each zone flow:

$$A_{cr}\frac{dQ_{cr}}{dt} = \Delta P_{core} - K_{cr}Q_{cr}\,/\,Q_{cr}\,/\,-g\int_{Z_{CE}}^{Z_{CT}}\rho_r\,dh, \, for\, r = 1\, to\, 10 \qquad (19\, to\, 28)$$

Total flow through the core is given by

$$\sum_{r=1}^{10}n_r\,Q_{cr} = \left(\sum_{l=1}^{2}Q_{PP1}\right) - Q_{L1} - Q_{Pi11} \qquad (29)$$

B.3 NUMERICAL SOLUTION

The equations describing the flow through IHX, pump flows, leakage flow, and sodium levels are solved by Hamming's Predictor corrector technique as outlined in Chapter 7. The total core flow obtained from the solution is used as input for the core flow zone equations. The solution of these equations is done by applying a semi-implicit finite differencing (Agrawal et al., 1977). By utilizing the fact that the total change in core flow obtained from the first stage is equal to the total change in individual zone flows, the calculation of the current time individual zone flows are obtained as follows:

$$\Delta P_{core} = \left[\frac{\delta Q_{RIj} + \sum_{r=1}^{10}n_r a_r b_r}{\sum_{r=1}^{10}n_r a_r}\right] \qquad (30)$$

$$Q_{crj} = Q_{cr\,j-1} + \Delta P_{core}\,a_r - a_r b_r, \, for\, r = 1\, to\, 10 \qquad (31\, to\, 40)$$

Where $\delta t = t_j - t_{j-1}$

$$\delta Q_{RIj} = \left[\sum_{l=1}^{2}Q_{PPlj}\right] - \left(Q_{L1j} + Q_{Pi11j}\right) - Q_{R1j-1}$$

$$a_r = \frac{1}{\dfrac{A_{cr}}{\delta t} + K_{cr}\left|Q_{crj} - 1\right|}$$

$$b_r = K_{cr} / Q_{crj-1} / Q_{crj-1} + \overline{d_r}$$

$$\overline{d_r} = g \int_{Z_{CE}}^{Z_{CT}} \rho_r \, dh$$

B.4 RESULTS

One-dimensional (1D) thermal hydraulic analysis was carried out using DYANA-P code with the normal primary hydraulics model replaced with the one for pipe break, to obtain transient evolutions of primary circuit flows and hotspot temperatures. Maximum temperatures are compared against the design safety limits (DSL). DSL for a Category 4 event are that the clad hotspot temperature (CHST) should be less than 1,473 K and SA mean sodium hotspot temperature (SASHST) should be less than the boiling point of sodium (~1,203 K).

The pumps start cavitating within 0.05 s and their flows increase to 126% instantaneously. Core flow reaches a minimum value of 39% of nominal at 0.6 s. It may be mentioned that in order to confirm the stable operation of the pump during cavitation, a 1:2.75 scaled model pump under severe cavitation (80% head drop) condition has been tested. The pump was operated for about 10 min and no flow fluctuation or vapor locking was observed.

The SCRAM parameters, its appearance time, and maximum values of T_{cl}, and T_{Na} reached during the event are given in Table B.1.

It may be noted that for a predicted fast coolant loss, one predicts an early SCRAM operation that is initiated by a sensing device, while with a slower coolant loss, the SCRAM may be initiated at a later time. Thus, it is important to employ a realistic rather than a conservative model to determine the discharge rate. However, the probability of such an event is low, and consensus is evolving to move this event to beyond design basis (IAEA-Tecdoc, 2003).

Table B.1 Maximum values of temperature reached for different trip parameters

SCRAM parameter and demand time	Maximum CHST, K (Limit 1,473)	Maximum SASHST, K (Limit 1,203)
P/Q at 0.06 s	1,267	1,107
Reactivity at 0.2 s	1,286	1,120
ln P at 0.45 s	1,323	1,144
Θ_{CSAM} at 1.1 s	1,398	1,201

REFERENCES

Agrawal A.K., Guppy J.G., Madni I.K., Quan V., Weaver V. and Yang J.W. (1977), "Simulation of transients in liquid-metal fast breeder reactor systems", *Nucl. Sci. Eng.*, Vol. 64, No. 2, pp. 480–491. http://dx.doi.org/10.13182/NSE77-A27384

IAEA-TECDOC-1406, (2003), "Primary coolant pipe rupture event in liquid metal cooled reactors", *Proceedings of a technical meeting held in Kalpakkam*, India, 13–17 January 2003.

Natesan K., Kasinathan N. and Vaidyanathan G., (1999), "Analysis of primary pump discharge pipe rupture incident", ICONE-7, Tokyo, Japan, April 19–23, 1999.

Index

Page numbers in *italics* refer figures and **bold** refer tables.

Actinides, 3, 4, 189
AHX *see* Air heat exchanger
Air Heat Exchanger, 35, 37, 203–211
Air Stack/Chimney, 209
Annular flow, 126
Annular Linear Induction pump, 102
Austenitic stainless steel, 49, 130
Axial flow, 90, 118

Beyond design basis events (BDBE), 165
Blanket, 2, 21, 43–45, 114
   lower, 32
   subassembly, 32, 61, 213
Blowdown, 27, 127–128
Boiling water Reactor (BWR), 1, 238
Bowing, 39, 51, 54
Breeding, 2, 32, 43
   Basics, 1
   ratio, 45
Bubble
   columnar, 223, 226
   growth, 221
   spherical, 221, 223, 225

Centrifugal pump, 25, 101–102
   Hydraulic model, 104–107
Clementine, 1, 6, 7, 47
Correlation
   Cheng-Todreas, 116
   fuel heat transfer, 75, 89, 139, 140
   nucleate boiling, 142
   pressure drop, 116, 145, 146
Coast down, 15, 30, 67, 86, 116, 156
   Duration, 163
   Law, 16, 104
   Time, 163, 165, 166, 169

Cold Pool, 35, 43, 66, 70, 207, 247
coolant
   distribution, 114
   flow, 23, 40, 115
   mixing model, 67–68, 71, 96–98
   selection, 47
   velocities, 41
Collier, 125, 141
Control and safety rods (CSR), 32, 182
   control margin, 178, 185
Core Thermal Model, 61, 152
   Validation, 65
Cover gas, 6, 24–25, 134
   Activity, 37
   expansion tank, 29
   Heat transfer, 62
   Pressure, 108, 113, 121, 218
CRBRP, 7, 8, 16, 68, 98, *164*
   Steam generator, 134
Cr-Mo steel, 132
Cross flow, 62, 80, 90
   Heat exchanger, 207
   Resistance coefficient, 117
   Velocities, 79

Dampers, 35, 192, 204, 207
Decay heat, 27, 35, 56
   curve, 56
   Exchanger (DHX), 35
   long term, 48
   removal (DHR), 9, *36*, 166, 189–191
Decay heat removal systems, 192
   FBTR, 194–195
   Operational Grade (OGDHR), *192*, *202*, 202–203
   PFBR, *202*

Primary Sodium, 191
  Safety grade (SGDHR), 191, 202
  Secondary sodium (SSDHR), 191, 192
Defense in depth, 10, 37, 104
Delay time, 177
  Instrumentation, 185
  Permissible, **186**
  Total, 182
Delayed neutron
  Detection (DND), 14, 178, 182–183
  half-lives, **51**
Design Basis, 10
  Events (DBE), 1, 11, 12, 37, 181
  Leak, 228–229
Design Safety Limits (DSL), 181, 250
DESOPT code, 147
DHDYN code, 204, 211
Direct current conduction pump, 102
Direct reactor auxiliary cooling system
      (DRACS) *see* safety grade
      DHR
Diverse safety rods (DSR), 32, 182
Diversity, 15, 172, 179, 183
  decay heat exchanger, 35
  Steam generator concepts, 133
Doppler effect, 52, 60
  Coefficient, 53
  feedback, 39, 51
Double ended Guillotine failure (DEG),
      228
Drum type, **134**
  boiler, 127, *128*
  steam generator, 128–131, *129*
Dry out, 126, **140**, 141, 169
  Post, **141**, 142
DYNAM code, 65, 91, 119, 151–154
  Organization, *153*
DYANA-HM code, 151, 155–160
  Comparison with DYANA-P code,
      156–161

Enrico Fermi Fast breeder reactor
      (EFFBR), 7, 180
Enveloping events, 11, 181
Evaporator, 31, 129, **147**
Event Analysis, 151, 170, 184
Evolution
  Core Flows, *160*
  IHX Primary Temperatures,
      157–158, 161
Experimental Breeder Reactor-I, 1
Experimental fast reactors, 7, 45

Flow
  blockage, 180
  bubbly, 126
  choked, 219
  halving time, 30, 35, 104, 165–172
  inter Wrapper (IWF), 212–213
  regime, 126, 197
FBTR, 7, 14, 19, *20*, 23
  Core layout, 44
  Flowsheet, 120
  Fuel subassembly, *46*
  SG modules, *194*
  ULOF in, *54*
  Upper plenum, *68*
Flywheel, 15, *25*, 101, 148, 163
Fuel
  axial expansion, 51, 53
  Mixed Oxide, 31, 45, 62
  Restructuring, 62–63
  thermal model, 63

Gap conductance, 63, 233
Gear's method, 118

Heat Transfer
  Data, 48
  Free convection, 76
  inter wrapper flow, 213
  Mechanisms, 125
  Primary system, 61
  Primary to secondary sodium, 90,
      212
  Reactor, 43
  Regimes, *126*
  Tube banks, 90
Hamming's Predictor Corrector
      Method, 118, 122, 236
Helical coil, 133, **140**, 224
Homologous, 248
  head -torque curves, 107, 108
  theory, 106
Hot Pool, 35, 66–69, 79, 181
  Hydraulics, 16
  Thermal model, 205
  Thermal stratification, 159–160
  Velocity patterns, 155
Hot Spot Clad, 65, 173, 179
Hydraulic model, 111–113

Inadvertent withdrawal of one control
      rod, 179, 184
Intermediate Heat Exchanger *see* IHX

IHX, 6, 15, 25, 111
  FBTR, 26
  loop type, *80*
  Pool type, *81*
  Pressure drop, 116
  Thermal model, 79

Kalimer-600, 17, 165, 174

LSSS, 12–15, 177, 185
  adequacy, 178
  evaluation of, **179**
  PFBR, 180
Labyrinths, 115
Limiting conditions for operation
  (LCO), 12
Limiting Safety System Settings *see*
  LSSS
lower plenum *see* cold pool

Nodal Heat balance Scheme (NHB),
  85–86
  Modified (MNHB), 86–89
Natural circulation, 15, 37, *122*
Natural convection, 16, 27, 35, 95, 116,
  203
  basics, 189
  flow, 122, 159, 167, 197, 213
Nucleate Boiling, 139, **140**

Orifice, plates, 47, 115–*116*
  Honeycomb, 115, *116*

Piping, 35, 38, 209
  coolant- wall (CW), 98
  coolant- wall- insulation (CWI),
    98
  heat losses, 122, 152, 195, 200
  schematic, *96*
  staggered mesh, 98
  thermal model, 95
PHENIX, 6, 8, 17, 55, 131, 193
  End Of Life Tests, 233
  Experience IHX, 79
  Steam generator, *132*, 133, 134
Primary pump, 12, 15, 33, 182
  coast down, 30, *119*
  discharge pipe, 113, 245
  inertia, 164
  seizure, 11, 184
  speed, 169, 170, 182
  trip, 11, 90, 180

Plant Protection system (PPS), 9, 12,
  37, 245
  FBTR, 177–180
  PFBR, 180–186
Point kinetics (PK), 49, 57
Pony motor, 35, 104–105
Post Dry out, 126, **141**, 142
Pressure drop, 45, 111–112, 142
  Casing SG, 196
  Coefficient, 16, 123, 196
  Devices, 115
  Fuel sub assembly, 116, *117*
  Two phase, 143
  Waterside, 138
Primary Hydraulics, 120–122, 151, 245
Prompt Jump Approximation, 57–59,
  236
Prototype Fast Breeder Reactor (PFBR),
  19, 31
  Cold pool, 70
  Core, 33
  Fuel, 114
  Grid plate, *75*
  Hot pool, 156
  Pressure transients, 227–229
  Reactor assembly, *34*
  Schematic, *32*
  Secondary sodium circuit, 227
  Shutdown system, 182–184
    Logic, 183
  Steam reheat, *133*
Prototype fast reactor (PFR), 6, 16, 104,
  174, 228
Pump tank, 109, 121, 199, 218,
  226–227
  Pressure, 107

RAPSODIE, 6, 19, 104, 172, 174
Reaction
  product Discharge Circuit, 218, 226
  sodium water, 16, 25, 121, 217–220
Reaction site, 125, 221–*222*, 228
  Dynamics, 218–221
  Pressure, 224
  Validation, 224
reactivity, 14, 28, 51–55
  feedback, 51
  structural expansion, 173
  total, 60
  void coefficient, 30
Reheater, 130, 134
response time, 14, 15, 55, 182

instrument, 9
    rupture disc, 228
    safety logic, 184
rod bundle, 115
    friction factor, 115–116
    heat transfer, 75
    wire wrapped, 116

Safety signals, 37, 178
Saturated boiling, 126
SCRAM threshold, 167, 169–171
Secondary pump, 35, 156
    drive, 148
    FHT, 171
    Inertia, 164
    trip/seizure, 11, 12, 72, 171
Secondary sodium circuit, 16, 19
    pressure relief system, 218
    pressure surge, 16, 121, 125, 218
Secondary thermal capacity, 213
Shutdown system, 15, 32, **174**, 178
    SDS, 182–186, **186**
Station Blackout, 11, 30, 156, 197
Steam Generator (SG), 6, 16, 125
    Casing, 30
    Designs, 127

FBTR, 27
Integrated, 31, 131
Leak, 29, 37
Material, 31
Module, 27
Once through (OTSG), 129
PFR, 133
SSC-L code, 16, 164
STITH-2D, 151–155
Superheater, 31, 127, 130, 134, 220
Super Phenix, 7, 65, 149, 193, 205, 224
SWEPT code, 224, 228

Trip Parameters, 186, 250

ULOF, 49, 172, 200
    reactivity coefficients, 172, 179
Upper plenums *see* hot pool

Water leak
    Inertia controlled model, 219
    Intermediate, 217
    Large, 29, 217
    Micro, 217
    Small, 217
    Stuttering flow model, 219

For Product Safety Concerns and Information please contact our EU
representative GPSR@taylorandfrancis.com
Taylor & Francis Verlag GmbH, Kaufingerstraße 24, 80331 München, Germany

www.ingramcontent.com/pod-product-compliance
Lightning Source LLC
Chambersburg PA
CBHW060350220326
41598CB00023B/2867

9 781032 254371